EDUCATION TODAY

Games in Geography

£1.00 net   20/-net

EDUCATION TODAY

For a full list of titles in this series see back cover

Games in Geography

REX WALFORD

Principal Lecturer in Geography and Senior Tutor
Maria Grey College, Twickenham

LONGMAN

GV
1485
W3

LONGMAN GROUP LIMITED

LONDON

Associated companies, branches and representatives throughout the world

© Longman Group Ltd 1969

First published 1969
Fourth impression 1972
ISBN 0 582 32084 4

Printed in Hong Kong by
Yu Luen Offset Printing Factory Ltd

Contents

INTRODUCTION   vii

1   GEOGRAPHY TEACHING : SOME PRESENT
PROBLEMS   1

2   NEW MOVEMENTS IN GEOGRAPHY TEACHING   13

3   GEOGRAPHICAL GAMING : A PRACTICAL ASPECT
OF MODEL THEORY   25

4   THE SHOPPING GAME   41

5   THE BUS SERVICE GAME   47

6   THE NORTH SEA GAS GAME   55

7   THE RAILWAY PIONEERS GAME   64

8   THE DEVELOPMENT GAME   77

9   THE EXPORT DRIVE GAME   85

10   TWO CLASSROOM EXPERIENCES OF
GEOGRAPHY GAMES   96

11   BUILDING GAMES AND SIMULATIONS   110
APPENDIX A; NOTES ON GAMES MATERIALS   117
APPENDIX B; SOME SUGGESTIONS FOR OTHER GAMES   120
BIBLIOGRAPHY   122

Above all, it is imperative for geography teachers to advance curriculum developments within their own subject, notably the use of decision-making games and the expression of geographical relationships in the form of models. What is wanted is not less subject teaching, but better subject teaching.

G. B. G. BULL

Geography, November 1968

Introduction

This text largely represents the working notes of an experiment that can be said to have only just begun. It ought not to be seen as a final or completely ordered piece of thinking, but as a help and guideline to those who may wish to take the arguments and intentions further in practical situations. It may help to reveal both the advantages and the limitations of a particular method in geography teaching.

The book had its origins in a small, duplicated pamphlet that was prepared in the Department of Geography at Maria Grey College of Education in 1968, following some initial experiments. Several teachers became interested and wrote to the college for information: and in order to further test the value of the games, the rules of them were set out, together with a brief introduction.

That original pamphlet has been much altered and extended but forms the core of this book. The six games described are now prefaced by chapters which try to set them within the wider context of changes in modern geography. The impact of such changes is now reaching some school classrooms. Following the games rules are chapters which describe two game experiences and which make suggestions concerning the development of games in both theoretical and practical aspects.

Game-playing is essentially a corporate activity and any value found in this book owes much to the co-operation and patience of those who have acted as willing partners in the developments. Teachers and pupils in several London schools have been helpful in their comments and criticisms, and several generations of Maria Grey geography students have cheerfully participated in much of the experimental work and invaluably aided it.

R.W.

I

Geography Teaching:
Some Present Problems

Teaching is, in some senses, a sheltered profession. . . . It has, in the enjoyment of learning, an especial temptation to dwell in the past, and even to feel at home in it. If the teachers bring the vitalities of the past to give life to the present all is well; but if their emphasis is such as to make their charges look backwards all is not well.

. . . Teachers represent the base of the pyramid of persuasion. . . . They may think that they are teaching subjects, but what, of course, they are doing, for better or worse, is something far more important in the long run. They are instilling an attitude to subjects.

JOHN GRIERSON
Grierson on Documentary

Geography in three schools

At one school which I once knew, geography was not noted for its important place in the curriculum. It was, in fact, taught only to the 'less able' in the forms up to 'O' level standard—and as a fill-up substitute for those who did not have the brains for Latin or Greek. There was not a full-time teacher of geography on the staff; the lessons were taken by a member of the P.E. staff when he had periods free from the gymnasium.

So far as I can remember no one at the school (pupils or staff) ever suggested that geography was anything other than a humdrum catalogue of descriptive material; on the one hand it dealt with the whereabouts of rivers and mountains on the earth's surface, and such things as what a moraine was; on the other hand it assembled information about the climatic requirements of such things as cassavas and assigned adjectives such as 'textile' or 'fishing' to settlements over a range of countries studied.

There was a general feeling abroad in the school that geography was pretty tame stuff. Bright pupils could gather what they wanted from atlases and encyclopaedias at home. Thus, you did not waste your time on geography if you could help it, but settled down to the more demanding tasks of experiments in the chemistry lab, or Latin unseens from the *Aeneid*. If you were going to university, you did not dream of doing geography—as a subject it was considered to have neither the guts nor the kudos that the rest of the curriculum had.

I know of another school in a small town not far from London— a grammar school now going comprehensive. Here geography is taught to most forms in the school and several pupils each year go on to university to read the subject. The calibre of pupils in this school is, perhaps, lower than in the first school. (Some unkind observers would equate the greater prominence of geography with that fact.) Geography in the sixth form is given three to four periods a week at minimum, with groups of fifteen to twenty pupils in each year. Four of the five 'O' level streams take geography, and examination results are satisfactory.

What is the nature of the subject in this school? It is geared to the reading of certain textbooks in strict chapter order, interspersed with note-taking from the board, and ten-question tests on the weekly homework. The pattern of work for the pupil consists in listening briefly to the teacher, copying notes (or sometimes a map) from the board and learning the notes and textbook in preparation for testing. Discussion is spasmodic since it holds up the ordered progress of the syllabus; project work is not considered an 'efficient' use of time. Participation by the pupils in classwork is almost non-existent. Their motivation to learn is the carrot of another 'O' or 'A' level pass dangled before them. Geography is given a prominent place in the curriculum and fulfils almost all the expectations of those who dismissed it as non-essential in the first school.

As a third example, I mention a school in which I taught for a spell. This was an average secondary modern school, where the pressure on children to take examinations was notably soft-pedalled, and where two periods a week of geography was the norm for all classes, from eleven to fifteen. I mention this educational institution not to cast stones at its general humane and liberal approach, but to acknowledge my own inabilities while teaching there, and to add yet another damp cartridge to this clutch of geographical misfires. I came to the school as a young supply teacher, the only geographer on the staff and one with no previous experience of what was required in the situation.

Casting about in the stock cupboard I found a great variety of textbooks, most of them dealing with parts of the world about which I knew little; I took them out and doggedly tried to work through a prescribed syllabus which included such areas as Afghanistan and Chile for study. The classes I taught included a sizeable proportion of delinquent boys in the upper forms; the general I.Q. figure was not high. I kept my classes quiet initially by dwelling on the more obscure sociological features of distant countries and feeling gratified at the wide-eyed open stares that were produced when we talked about (say) polygamy or cannibalism; the technique on which I also heavily relied was that of tracing maps. It seemed to be an ideal child-minding device and was a needed respite while I looked up the book and tried to find what came next. Thus teacher and pupils learnt together.

These three situations are not, thank goodness, entirely representative of geography teaching. But they do represent, perhaps more frequently than one would like, the kind of approach that gains geography low repute.

Lack of preparation time, inadequate equipment, oversized classes, the pressure of examinations, all contribute to exacerbate the situation. Even the dedicated enthusiasm of the highest order knows what it is like to try to take forty fourteen-year-olds on a rainy afternoon in a Victorian classroom without proper atlases available, and teach them, say, something about the Ruhr for G.C.E.

But beyond these technical problems lie two others at the heart of some current geographical despair:

(a) the concern about whether the curriculum which we teach is suitable

(b) the suspicion that our techniques of teaching may need variety and revision.

Curriculum and its problems

Curriculum building is usually such a lengthy and protracted activity (particularly when further fossilized by the statutes of examination boards) that it inevitably runs a grave risk of being constantly one step behind the needs of a fast-changing society. In no subject is that more likely to be true than in geography.

It is not only *facts* which may be a devalued currency. The *ideas* behind the structure of curricula may themselves be outpaced; they may be inadequate because of their compartmentalization, or because of their lack of depth, or for a score of other reasons. It is not fair to blame curriculum builders of the past for these inadequacies; the blame attaches more to the generation which fails to recognize the need for development and evolution in the curriculum and makes do with yesterday's clothes in fitting pupils for tomorrow's world.

If one looks at geography as it is currently taught in many schools, I think it not too harsh to say that it probably suffers from at least one of three defects (perhaps more than one, or even from all three).

IT MAY BE SUPERFICIAL

Ploughing our way from the exports of Venezuela, to the temperatures of the Amazon basin, we teach an assemblage of facts about South America—perhaps enough to persuade the G.C.E. or C.S.E. examiner that the pupil knows Bogota from Brasilia or Bahia from Buenos Aires. But despite the attempts that we or the textbook may make to examine problems critically in passing, in the mind of the pupil geography work on South America remains little more than a loose jumble of material— learnt for the examination room, and jettisoned directly after it has been used there. It seems to have no place in what is considered by pupils as 'worth remembering' either for later use in life, or even for interest's sake.

I recently asked a nineteen-year-old friend of mine just what he *did* remember from his four years' geography at school. He cast about desperately for several minutes before finally saying 'It's no use—it was just a set of names and things that we learnt—we never really *understood* anything about the countries'.

H. G. Wells's Kipps, you may remember, was a little more confident than my friend. When the lovely Miss Helena Walsingham asked him if he knew any geography, he was confident that he had 'done geography'. His recall efforts brought him 'Tyne, Wear, Tees 'Umber', (The first four rivers on the map of England and Wales, if learnt clockwise), and 'Bickley, Bexley, Bromley' (three stations on the London to Folkestone railway line).

Though no doubt it is pleasant to have this kind of information at one's command, it is hardly more than disconnected fragments, strings of names and dates, and perhaps sets of figures trotted out with cliché phrases that parrot the textbook or the teacher's notes. It is an indiscriminate confetti of material that remains only on the surface of the memory because of its lack of structure.

Because of this lack of structure, it not only falls away from recall quite sharply—it also fails to penetrate to the heart of the subject.[1] In this way, gross over-simplifications are resurrected to the status of Geographical Commandments, and statements are transformed into shibboleths because of lack of scrutiny given to them.

Pupils and teachers are not the only conspirators here. So often an examination perpetuates superficiality by testing at a level which demands only facile recall. The ability of a pupil to understand in depth is rarely susceptible of being properly examined.

IT MAY BE IRRELEVANT

Superficiality and irrelevance are likely to go close in hand— at least at one level. The resemblance between Kipps's knowledge, and that which N. F. Simpson's Prosecutor (in *One way pendulum*) expects from Arthur Groomkirby is apparent.

PROSECUTOR: There must have been quite a number of places from which you absented yourself on that rather vital twenty-third of August, Mr Groomkirby, in order to be in Chester-le-Street?

ARTHUR: I dare say that would be so, yes sir.

PROSECUTOR: You were not, for instance, in London?

ARTHUR: No sir.

PROSECUTOR: Or Paris?

ARTHUR: No, Sir.

PROSECUTOR: Or Rome?

ARTHUR: No, I wasn't there, sir.

PROSECUTOR: You were not, I imagine, in Reykjavik, either?

ARTHUR: I couldn't say for sure where that is, sir.

PROSECUTOR: Yet you absented yourself from it?

ARTHUR: As far as I know I did, yes.

PROSECUTOR: *And* from Kostroma.

ARTHUR: I suppose I must have done.

PROSECUTOR: And Chengtu, and Farafangana, and Pocatello?

ARTHUR: I'm afraid I'm not all that much good at geography.

PROSECUTOR: Not much good at geography, Mr Groomkirby— yet you want the Court to believe that in order to be present at Chester-le-Street you absented yourself from a whole host of places which only *an expert geographer* could possibly be expected to have heard of?

Besides Chengtu, Farafangana and Pocatello, I find that I, as a geographer, am also often expected to know such things as the population of Alaska (exactly) or how many counties there are in the United Kingdom (the latter an enquiry recently made by phone from the managing director of a plastics firm). If we geographers taught our pupils more emphatically *how and where to find these things out* should they be needed, we would have a little more time, perhaps, to deal with the central themes of our subject.

But our irrelevancies stretch over more than exotic place names. We also survey the picturesque (perhaps in desperation) in order to make the subject more interesting, we hope. The obscurities of social anthropology in the South Pacific, or the occasional tragic accidents of nature are surely not the main aims of geographic education—yet this kind of detail (though admittedly of marginal interest) is usually all that is remembered from classroom teaching in some schools. Infinite detail about how the Gilbert and Ellice Islanders catch fish with their feet, or how Congo pygmies bring up their young to face a nomadic existence is remembered to the exclusion of any kind of comprehension about the factors that locate industry, or the importance of the Warsaw Pact on the geography of Europe. Bombo of the Congo has had a long and well-loved reign (particularly in the primary school); but he, together with Anton of the Cantons, Abdul the shaduf boy, and others of that ilk, should be honourably retired to the limbo of folk-tale where they now, for the most part, belong. Geography in schools should not be bound up at any level with the preservation of the picturesque in antique societies.

There is a third kind of irrelevance. It is the irrelevance engendered by the Cook's tour approach of most curricula, which invite us to race round the world two or three continents a year at the start of the secondary school—that gives us forty minutes for Peru and sixty minutes for Zambia (if we are lucky)— and gather a few facts as we go. The superficiality of the information is paralleled by the general uselessness of the emasculated descriptive regional survey which is usually the end product of such a tour. The great synthesizing work of Vidal de la Blache and the French geographers has now become transmuted into a guide book approach to world geography which we do as a kind of reflex action, because we believe that regional survey is 'what geographers are supposed to do'. There is strong reason to believe that Vidal himself observed the shortcomings of his own methods,[2] but we have gone on compiling dutiful regional descriptions until shortcircuited by the approach in depth of other disciplines.

Regional survey has, I believe, led us in the past into ignoring some of the realities of the world situation. Some problems were left alone because they were not 'geographical'; others were ignored in order not to break up the 'geographical unity' (supposed or imagined) of an area. But the finely distilled regional appreciations of expert geographers were often overlain on the ground by rather more prosaic considerations, and some of the hallowed geographical deductions made in the name of the evident relationship between man and landscape have lately been exploded.[3]

Rather than regional survey, it may at present be better to concentrate on topic areas or particular problems—selected almost by their relevance and topicality alone. Growing enough food for the world to live on, or holding China in check can be *seen* by pupils to affect their future lives in an immediate way; an orthodox commodity study of rice, or even of South-East Asia as a region may not have the same urgency, or ultimate relevance.

IT MAY BE OUT-OF-DATE

Marshall McLuhan, perhaps more than anyone else has pointed to the changing nature of the contemporary world; his assertion of the re-creation of the 'global village' is not simply a communications image—it is an intensely geographical one.

Life at the Pentagon has been greatly complicated by jet travel, for example. Every few minutes an assembly gong rings to summon many specialists from their desks to hear a personal report from an expert from some remote part of the world. Meanwhile the undone paper work mounts on each desk, and each department daily despatches personnel by jet to remote areas for more data and reports. Such is the speed of this process . . . that those going forth to the ends of the earth often arrive unable to spell the name of the place they have been sent to as experts.[4]

Geographers may recognize at least an echo of their plight in McLuhan's picture of the Pentagon administrator. The turning of the world into the global village produces not too little data, but too much.

A hundred and twenty years ago we could still write 'INDIANS' across a map of the western part of the United States in the knowledge that this was almost the only factor of the human geography of the area of which we had knowledge. In 1967

J. P. Cole and G. A. Smith, in a duplicated bulletin of less than twenty pages, produced sets of information about American states[5] and pointed out that the possible (sensible) combinations of these variations in some form of factor analysis was roughly equivalent to the number of atomic particles in the universe.

To add to our troubles, this mass of information is not only changing, but growing—at a rate equivalent to the duplication of the brooms in *The Sorcerer's Apprentice*. Our knowledge is doubling every fifteen years, it has been estimated. Several of the contributions in *Geography's* feature 'This changing world' are themselves out-dated in some respects by the time they appear in print. Plans for development are changed, market factors alter, population growth accelerates, communications develop— and all, it seems, at a speed of bewildering rapidity.

Yet the clammy hand of tradition still grips some of the work that we geographers do, even though the life-span of textbooks now approaches the obsolescence rate of thin aluminium frying-pans in Detroit—purchased to be used once and then thrown away. This despair about keeping pace with change in factual material may prove to be a blessing in disguise; but it is chastening still to walk into classrooms and find (as I swear I did in February 1968) children using textbooks which were printed in 1934. (On one page was a fine picture entitled 'Modern New York'; it was dominated by the shape of the Graf Zeppelin hovering prominently next to the Empire State Building.)

Thirty-five years ago Fairgrieve was urging geographers to turn their attention towards the 'world stage'. The description carried with it a connotation that the action unfolded in a fairly coherent and unhurried way, so that watchers could readily follow and analyse what happened on the stage. I hope it is not irreverent to suggest that Fairgrieve's stage is today transformed into a screen for a psychedelic light-show— a dazzling and overpowering kaleidoscope of events, with a tempestuous rate of change that races too fast for the human mind to fully comprehend. We try to abstract objective material from it for study, but the pattern will have changed again before we finish our task and so we take refuge in material already well known, whether relevant or not.

This total situation was recently observed by the four Council of Europe conferences which reviewed Geography Teaching and the Revision of Geographical Textbooks and Atlases. B. S. Roberson, reviewing the collected papers of the conferences (edited by E. C. Marchant) noted that 'Common complaints are that the picturesque is described at the expense of the normal, that primitive and disappearing ways of life are still given prominence, that there is still too much determinist emphasis, and above all, that economic pictures are out of date'.[6]

Teaching techniques

These problems of superficiality, irrelevance and being out-of-date are concerned with the subject-matter of geography; but there is another problem too. Whatever we are teaching, it seems dangerously prone to the accusation that it is just dull and boring. In a 1967 examination report a chief examiner wrote: 'It is disappointing that only a minority of the candidates . . . convey a sense of enthusiastic interest in geographical work.' Other sources suggest a similar disillusionment.[7]

One hundred and thirty students in a college of education year group, asked to evaluate their school geography teaching, placed 'Dull' at the top of their popularity poll when choosing suitable adjectives. Some may say that school work should necessarily be dull in parts; but this argument applies better to subjects other than geography. If the present environment of man cannot be made stimulating, challenging and exciting to citizens of the future, what possible hope is there for us all? We may as well pack our bags and leave the classrooms empty or install TV in every room and leave the mass media to get on with the job.

If modern geography teaching is too often dull, then one suspects that our techniques of teaching the subject must bear some part of the blame. The 'take-down-these-notes/learn them/here-are-ten-questions-to-test-you' syndrome operates emphatically in many classrooms, either because teachers feel ill-equipped to tackle anything else, or because they yearn for a quiet life or a safe examination result. In places where less able children do not respond to such tranquillizers, the tracing doodle or the 'draw a picture' dessert-course assumes a ritualized role. They can be predicted by the pupils with an ease approaching infallibility. (One excepts from this the country school in which a teacher working with remedial classes told me with some pride that he had found an answer to stimulating interest— 'I run films backwards and they can't take their eyes off the screen. They're as good as gold.')

The problem of non-motivation probably stems from two sources. On one hand we need to make pupils participate more in lessons—asking questions, venturing judgments, assuming responsibilities. On the other hand, we need to vary our techniques of teaching much more, so that geography lessons are not so utterly predictable. New approaches to the teaching of the subject should not seek to overthrow all the established techniques overnight. But they should be given a chance to make their way and add to the variety of possibilities from which a teacher may choose.

Mayo's work in Illinois[8] established that what he called 'the Hawthorne effect'—the innovation of new procedures, was itself a spur to learning whatever the value of the procedure introduced; and new ideas in our classrooms may often have a similar effect in shaking a class from its customary lethargies. What follows in later chapters of this book is simply an attempt to introduce one new technique and to relate it to some of the changes now taking place in the way in which geography is being considered as an academic subject. These changes, too, have much bearing on what goes on in the classroom.

References

1 BRUNER, J. S. *The process of education* chapter 2, Random House, 1960.
2 See WRIGLEY, E. A. in *Frontiers in geographical teaching*, pp. 7-13, ed. R. J. Chorley and P. Haggett, Methuen 1965.
3 See, for instance, WATTS, D. G. 'Changes in the location of the South Wales Iron and Steel Industry, 1860–1930', *Geography*, No. 240, **53**, part 3, July 1968. Watts comments that the changes were due principally to human and chance factors, illustrating well how these can overcome physical handicaps.

4 MCLUHAN, MARSHALL, *Understanding media*, p. 59, McGraw-Hill, 1964.

5 COLE, J. P, and SMITH, G. A. *Bulletin of quantitative data for geographers*, No. 8. University of Nottingham, Dept. of Geography, 1967.

6 *Geography*, No. 243, Vol. 54, Pt. 2, April 1969.

7 See, for instance, PARTRIDGE, JOHN, *Life in a secondary modern school*, Penguin, 1966, especially chapters 6 and 8. Also CAVE, RONALD G. *All their future*, Penguin, 1968.

8 MAYO, ELTON, *The social problems of an industrial civilization*, Routledge, 1957.

2

New Movements in Geography Teaching

All change tends to be applauded when it happens to someone else in history books, but heterogenetic change, which involves a complete break with the past, is apt to be very painful to those involved in it. . . . Really significant revolutions in ideas, however, stem not from relatively minor changes carried out within the conceptual framework of the age, but from . . . what Thomas Kuhn would call the creation of a new paradigm.

It is a change of paradigm which I believe to be taking place in geography. . . . Of course [it] will not only affect research. The new categories will have repercussions throughout the geographical hierarchy. . . . [It] will presumably eventually induce associated modifications in that last bastion of geographical conservatism, the school examination boards, and, finally, in the school classrooms themselves.

PAUL WHEATLEY,
Professor of Geography,
University College, London.

Some new developments

The state of affairs described at the beginning of the previous chapter has concerned geographical educators in various ways.

The liberation of primary school work from deskbound formality was achieved comparatively early in Great Britain, following the Victorian and Edwardian periods, and in many junior schools today children are more frequently out of their desks than in them. The success of this method has been charted effectively[1] and British primary education, under the influence of such figures as Susan Isaacs, has been accounted a leading example of progressive development. In geography, Olive Garnett's classic text in this sphere was published as early as 1934, and still remains a valued addition to any education bookshelf.[2]

In the secondary school, despite the cramping restriction of examinations, there has been experiment in recent years; the recent institution of the Schools Council follows a growing concern about curriculum and method in many subjects, and a desire among teachers to control more of their own destiny. In geography, the work of men like G. J. Cons and R. C. Honeybone has helped to breathe more reality into syllabuses and activities. The most recent book to witness to a continuing development in the mainstream of geography teaching has been B. S. Roberson and I. L. M. Long's *Teaching geography*.[3]

The role of the Geographical Association, the organization to which most geography teachers belong, has been important. Its own publication about junior school work[4] provided sound guidelines for those who were prepared to dip into its slim covers and gather from its accumulated wisdom; in secondary schools it encouraged the use of sample study material[5] to bring geography alive through more particularization, and emphasized the value of local study and the understanding of maps as an important geographical skill.[6]

While this has been happening, there have been radical changes developing in other subjects. The orthodox computation of old-style arithmetic and algebra has become transformed by the swinging image of the 'new maths'; the development of 'Nuffield Science' has similarly sparked off interest and excitement in that branch of the curriculum.

And in geography also, related to the above developments, there has lately been a radical ground-swell, dubbed by some, rather doubtfully 'new geography'. Its freshness is not in any reshuffling of subject matter, but in its desire to create a distinctive geographical methodology. It is born, one feels, out of unhappiness with *lack* of methodology, and out of a desire to bring together more firmly the disparate elements that descriptive synthesis has struggled so desperately to encircle in the last fifty years. Its concern with methodology has, fortunately, also liberated the emphasis on 'covering everything' which previously lurked at the back of most syllabus construction; its supposition is that method in depth may mean less actual subject material covered in school—but more ability to do so subsequently.

Professor Wheatley of University College commented on this development recently.

> I think that there is one development of first importance taking place in British geography at the present moment [1968] . . . hitherto the scatter of alleged geographical interests has been, in my opinion, one of the chief impediments to progress. . . .
>
> It is my impression that now a new unity is beginning to emerge . . . my guess is that [this new and conceptual framework] will be fairly closely related to that bundle of cognizances and skills which are coming to be known as locational analysis. The editors of a recently published book have proposed the model as a likely means of integrating hitherto disparate geographical traditions. They envisage, in their terms, a model-based paradigm replacing the old predominantly classificatory geographical tradition. . . .
>
> I agree one hundred per cent with the thesis of this work in its implicit assumption that geographical study at all levels is in need of, and is fortunately being provided/with, new categories of thought. And this means a break with the past.[7]

The development of which Wheatley speaks began significantly in Britain as late as 1963. Between that year and 1966 a set of seminars was held for teachers at Madingley Hall, Cambridge, under the auspices of the University of Cambridge Extra-Mural Department. Lectures from two of these Madingley seminars were later published under the titles *Frontiers in geographical teaching* and *Models in geography* and these two books (the latter referred to by Wheatley above) have already taken their place as seminal handbooks to the early stages of the new developments.[8]

The Madingley conferences and books brought together for the first time strands of thought and investigation that had been developing in some university departments and wove them into an exciting and provocative mixture of challenge to the existing tradition.[9] The organizers of the conference and editors of the books, Mr R. J. Chorley and Dr P. Haggett (both, at that time, of the University of Cambridge), were themselves key figures in these developments.

The ripples that began to develop from these Madingley conferences soon caused reactions. A group of teachers in London formed a Madingley-inspired organization called the London Schools Geographical Group and found, to their surprise, that four hundred teachers turned up to their first tentative day-conference (the programme of which had hinted at the new ideas abroad). Geographical Association branches began to discuss the ideas and magnificently partisan feeling arose on both sides to enliven and invigorate their programmes. At national level, the G. A. felt constrained to set up a separate executive committee to evaluate and consider the new developments[10] to consider the topic in several sessions at its national conferences and to devote a whole issue of *Geography* to articles written in a 'Madingley'[11] vein.

Attention was drawn to similar developments in the U.S.A., where a government-aided research project was translating similar university work into classroom terms. The 'American High School Project', as it has become popularly known, is not yet complete, and its literature is hard to come by, since it is not generally released; nevertheless it has been chronicled in Britain[12] and its influence has been considerable—even through duplicate and photostat copies of fragments of the original documents.

There was also a subsequent consideration of parallel developments in other fields. The relationship of 'new maths' and science has already been mentioned. But geographers were also considering what was happening in such fields as business management, economics, political science, town planning and sociology. It was encouraging to find that the ideas and techniques being developed in these related fields harmonized well with 'new geography'; the almost deliberate vacuum which lay between traditional geographers and other subjects has been easy to fill as integrated study of problems has developed along fresh lines.[13]

All this has happened within a five-year span, so it is impossible to evaluate or properly place in perspective these new developments. Nevertheless there is every sign that they are beginning to have their impact in the classroom.

To use the term 'new geography' in indiscriminate fashion is dangerous, and it would be unfortunate if some were daunted by its pretentiousness. Hardly anything is ever really 'new'; what is described in this vein is usually a revival of a long-forgotten or misunderstood truth which had been buried by time and academic distance. What is 'new' about 'new geography' is only the fact that it states with some freshness past truths that some of the classical geographers took for granted.

But it is certainly born into a largely sceptical world. Chorley and Haggett identified three problems in the 'academic log-jam' that seems to have held up progress in geography at university level in the past sixty years, and with these problems they have sought to grapple:

1 a 'Victorian academic structure every bit as solid and constraining as its architectural counterpart—the everyday popular image of geography is an antique one dogged by "exploration description and capes and bays".'

2 conflicting objectives. 'Much of its popularity springs not so much from its possessing any satisfying basic academic discipline as from "valuable side-products" which are believed to spring from its study.'

3 academic isolationism. 'There is no doubt that the most sterile aspects of present geography are the result of academic inbreeding.'[14]

These criticisms of academic geography need to be laid alongside those made of classroom geography in the first chapter. If founded in truth, together they give cause for concern, if not for alarm.

The core of new approaches

What is it then that the 'Madingley' approach has sought to emphasize? What contemporary images, satisfying academic disciplines, virile aspects of the subject can it produce? Much of its work, certainly, is tied up with a *re-emphasis on the geometrical aspect of geography*. Chorley and Haggett see the geometrical tradition as basic to the original Greek conception of the subject, and one on which we have concentrated too little in recent years. Geography as 'the geometry of the real environment' has underlain much of the creative research work of the past (they quote Christaller and Wooldridge in the same breath in instancing this unity), and needs to be seen as such again. 'Geometry not only offers a chance of welding aspects of human and physical geography in a new working partnership, but revives the central role of cartography in relation to the two.'[15] A recent book edited by Berry and Marble has illustrated this point vividly,[16] in bringing together research papers with a common bond in their approach to spatial analysis.

As Wheatley indicated, this geometrical approach to geography sees the subject primarily as the science of location in space, and a subject concerned with the spatial characteristics of points, areas, networks and so on. This interest and concern for spatial theory not only leads to much deeper and analytic study of maps and diagrams (in contrast to the rather inadequate scanning and verbal description of them in the past) but also importantly liberates geography from the 'man and the land' equation which has sometimes affected it in a neodeterministic way.

Is the study of geography one of distributions and locations — whether it be distributions of potatoes, or patrons in a cinema, whether it be locations of woollen mills or washbasins in an office-block? Are the routing problems of a refuse cart as geographical as those of a railway system? It is round this which debate centres in one field. The 'new geographers' suggest that it is the methodological approach which is the distinctive mark of the subject, and not simply its traditional subject matter; they believe that investigational techniques can be applied over ranges of problems not previously considered the province of the geographer, and also applied analogously in quite different branches of his traditional discipline.

These geometrical approaches are allied to two other important strands in the ideas that have developed. One is the *increasing use of quantification and statistical analysis* to solve problems. The increase in quantification, though innovative (and dismaying) to some geographers, is only parallel with what is happening in other disciplines. The increase in data available through official sources has often been in tabular form, and the problem has increasingly become not where to find information but how to disentangle it.

The subjectivity of regional description has been buttressed by the inadequacy of words; but 'large', 'major' and 'some' have become increasingly less useful as descriptions of settlements, crops, and industries. The approach of the wine-taster to the subject has become invalidated even more when crude interpretation of statistical material has swept away popular misconceptions about topics.

It has also become apparent that by the use of numerical measures, more sophisticated work can be done. Chorley emphasized this in relation to geomorphology[17] in a vigorous chapter in *Frontiers in geographical teaching*, and in the same book D. Timms did likewise for work in urban studies.[18] Cole later produced a geographical account of France based on correlations.[19]

These numerical approaches have increased considerably in recent geographical output, though some geographers have been concerned about their own inability to cope with the maths involved. It is true to say, however, that some of the calculation looks far more forbidding than it actually is—in many exercises a level of computation not beyond that of a fifteen-year-old is demanded.

A considerable literature now exists on this topic—notably in America.[20] Burton's well-known article in *The Canadian Geographer* usefully outlines the history of the 'quantitative revolution' which established itself transatlantically in the 1950s.[21] He carefully notes the opposition to the idea, and its consequences for geography; he is able to say 'the revolution is now over, in that once-revolutionary ideas are now conventional'. The statement might yet be contested in Europe, but the struggle is on.

The third strand of the Madingley approach (and the one mentioned earlier by Wheatley) was developed to the full in the second book of lectures which was published, and which was also broken down into three smaller paperback volumes.[22] This *emphasized the importance of models to geographers*—recalling the traditionally used 'models', but at the same time pointing to their inadequacies. Used in this sense, the term includes both the 'hardware' variety and those of a theoretical kind.

The idea of the model is bound up with the two strands mentioned earlier, but represents probably the central strand of fresh thinking. It draws attention to the importance of *nomothetic* rather than *ideographic* teaching, i.e. towards the idea of looking for similarities rather than uniquenesses. (In this respect, it is counter to the method of regional survey of the French school.) Chorley and Haggett provide an acceptable and probably influential definition of a model in the first chapter of *Models in geography*.

Models can be viewed as selective approximations which, by the elimination of incidental detail, allow some fundamental, relevant or interesting aspects of the real world to appear in some generalized form.[23]

They are, in practice, chains of ideas linked together and observed at work; basic 'skeletons' which can be made to 'dance' at the command of the observer. One well-known recent example is a model devised by R. L. Morrill who sought to account for the evolution of the pattern of towns in Sweden.[24] Using:

1 Knowledge about original growth points.
2 Observed relationships between migration and settlement size.
3 Knowledge about developments in transportation.
4 Calculated possibilities of chance events.

Morrill simulated the evolutionary patterns in simplified form and used the model to suggest possible lines of development in the future.

The suggestion that models should be frequently used and investigated in geographical teaching carries with it the likely assumption that cursory inspection is unsatisfactory. A model needs to be looked at properly if at all.

Consequently the teacher who chooses to consider models

important will probably also be unable to cover the same amount of curriculum material as one involved in orthodox teaching. The attempt to 'race round the world' in a five-year scheme will not be practicable.

The crucial question will be whether the study of particular models equips a pupil better geographically, compared with the existing methods of study.

Some of the possible advantages in the use of models are set out below:

1 They may help to produce a *deeper understanding* of complex processes, because of the elimination of 'background noise'. The model simplifies the real world in order to abstract *fundamentals* for study and observation.

2 They may help to make *more sense of the real world*, because the latter presents a confusing first-face that is often a deterrent to close examination (e.g. the multiplicity of statistics available about a country may well dissuade a pupil from attempting to understand even the vital ones).

3 They may help to produce *better 'retrieval' in the memory*. Recent educational psychology has proved that the more structured thoughts are, the easier the task of retrieving them at a later date. Past geographical work has often notably lacked such structure.

4 They may assist in the *transfer of essential knowledge from one situation to a similar one*. In a world changing as fast as ours, the understanding of basic processes is likely to be more useful than the knowledge of actual facts, for these may be ephemeral. Therefore the study of a model (say) of a developing country may yield basic understandings which can be applied later in several different areas of the world; whereas a superficial knowledge of these countries is likely only to be of marginal value in the study of their real problems.

To return for a moment to the metaphor of the world stage and its current metamorphosis into a light-show, suggested in Chapter 1; it may now be better to concentrate our attention, not primarily on the dazzling patterns themselves, but on the machinery that controls and initiates them. Besides providing relief for the eyes, it is a somewhat simpler job. If we can understand what makes the patterns—even in simplified outline—then we can understand what is happening on the screen in a much more informed way. The screen is put in better perspective, as the machinery is understood. With knowledge of this, a subsequent light-show may be understood—or even predicted. Model users and builders would suggest that this ability to predict would be a useful adjunct to the understandings usually conferred by geography. It is not a usual characteristic of present approaches. To conclude the metaphor—back in the darkened room, a present bemused audience which chooses to keep its eyes glued *only* to the light-show on the screen may only have a succession of partial and uncomprehending experiences which are transitory in the extreme.

It is true, of course, that some 'models' have been in use for many years (the Davisian river cycle is a striking example of one that has long been used in an expository way in schools), but there have been very few in any human geography. To suggest that they become a key part of the way in which geography is studied is a major break with tradition for many classrooms.

The accelerating rate of change in human activity may be the crucial factor which eventually weans many away from the chase of encyclopaedic fact-gathering. Where facts get out of date so fast, the understandings of process are the more constant. In this sense, Tom Lehrer's comment on new maths: 'It's much more important to know what you are doing than to actually get the right answer', may be conceived in satire but rooted in truth. In much of modern geography many of the answers are changing too fast for them to be *worth* knowing.

The trend towards these new ideas meets in some measure the criticisms of curriculum voiced in Chapter 1. Superficiality and irrelevance are replaced by deeper study of more restricted material; being out-of-date is largely irrelevant with a change of goals in which some processes are considered to be more important than facts (though one would wish here *not* to deny the need for a basic framework of factual material on which this later study is based).

But is this kind of geography still dull? Certainly some think it fairly forbidding and strong meat for the weak of stomach. There seems to be a belief that 'new geography' may be all very well for the high-flyers but will founder when it is taught to the less able. One hopes that this is not so. There seems to be every reason why the use of models, for instance, should be less dull than some of the more orthodox teaching techniques in the classroom. Class participation is essential in the development and evaluation of models, and may well also be used in the building of them. Models offer more scope in the normal classroom situation; and a real chance to bring motivation to classes, of all abilities.

What follows in later chapters of this book is a minor attempt to demonstrate this point in a practical way.

References

1 GARDNER, D. E. M. *Experiment and tradition in primary schools*, Methuen, 1966.

2 GARNETT, O. *Fundamentals in school geography*, Harrap, 1934.

3 ROBERSON, B. S. and LONG, I. L. M., *Teaching geography*, 1967.

4 *Teaching geography in junior schools*, Geographical Association, several editors, new edition 1970.

5 *Sample studies*, Geographical Association, several editors.

6 Notably through its *British landscape through maps* series, which provides supplementary information and interpretation to 1 inch Ordnance Survey maps.

7 WHEATLEY, P. in *The Bloomsbury geographer*, University College, London, 1968, pp. 6–7.

8 CHORLEY, R. J. and HAGGETT, P. eds. *Frontiers in geographical teaching*, Methuen 1965, and *Models in geography*, Methuen, 1968.

9 See, for instance, the conflicting views of *Frontiers*, reviewed in *Geographical Journal* **132**, Pt. 4, Dec. 1966, p. 572 and *Geography*, No. 234, **52**, Pt. 1, Jan. 1967, p. 108–109. See also, the conflicting reviews of *Models in geography* in *Geographical Journal* **134**, Pt. 3, Sept. 1968, p. 405. and *Geography* No. 241, **53**, Pt. 4, Nov. 1968, p. 423, and *Geographical Review*, Vol. 58, No. 4, Oct. 1968, p. 674–5.

10 The Standing Committee on the role of models and quantitative techniques in geographical teaching.

11 *Geography*, No. 242, **54**, Part 1, Jan. 1969.

12 GRAVES, N. J. 'The High School Project of the Association of American Geographers', *Geography*, No. 238, **53**, Part 1, Jan. 1968.
The High School project issues newsletters from time to time, and other materials. Its address is c/o P.O. Box 1905, Boulder, Colorado, 80302, U.S.A. Much of its experimental work is continuing.

13 See the development of Regional Science both in the U.S.A. and in

Britain; several of Britain's newer universities have chosen to create Chairs in Environmental Studies, rather than in Geography.

14 *Frontiers in geographical teaching*, Chapter 18.
15 *Frontiers in geographical teaching*, pp. 376–7. Also, Haggett, P. *Locational analysis in human geography*, Arnold 1965, pp. 9–16.
16 BERRY, B. J. L. and MARBLE, D. F. *Spatial analysis*, Prentice-Hall, 1968.
17 *Frontiers in geographical teaching*, chapter 8.
18 *Ibid.* chapter 12.
19 COLE, J. P. *Bulletin of quantitative data for geographers, No.* 1, University of Nottingham, Dept. of Geography.
20 GREGORY, S. *Statistical methods and the geographer*, Longmans 1966, and COLE, J. P. and KING, C. A. M. *Quantitative Geography*, Wiley, 1969, are important British contributions to the topic.
One of the most useful introductions is Yeates, M. H. *An introduction to quantitative analysis in economic geography*, McGraw-Hill, 1968.
21 Reprinted as chapter 1 of *Spatial analysis*, originally published in *The Canadian Geographer*, **7**, 1963, pp. 151–162.
22 CHORLEY, R. J. and HAGGETT, P. eds., *Socio-economic models in geography, Physical and Information Models in Geography, Integrated models in geography*, Methuen, 1969.
23 *Models in geography*, p. 12.
24 MORRILL, R. L. 'The Development and Spatial Distribution of Towns in Sweden', *Annals of the Association of American Geographers*, **53**, pp. 1–14, 1963.

3

Geographical Gaming: A Practical Aspect of Model Theory

Games, like art, are a translator of experience. . . . The world of science has become quite self-conscious about the play element in its endless experiments with models of situations otherwise unobservable. . . . Galbraith argues that business must now study art, for the artist makes models of problems and situations that have not yet emerged in the larger matrix of society, giving the artistically perceptive man a decade of leeway in his planning.

MARSHALL MCLUHAN
Understanding media

The justification for the greater use of models in the classroom lies not only in their importance to research workers in the universities or in their centrality to curriculum improvement; it lies also in the possibility of motivating children better through their active participation in model-building, model-operating, and model-evaluation.

The use of the model in the classroom is not, of course, simply a magic wand to wave over geographical pumpkins in order to transform them beyond recognition—if introduced didactically and lectured about, in the same way that traditional material is used, it may produce no effective change in interest or motivation.

But there does seem to be opportunity for participation and involvement in models at a realistic level. And where classes themselves have been part of the model in its operation, interesting results have occurred. If the model is given something more than just a cursory glance, its properties begin to attract and hold attention.

Classification of models

It may be necessary, at this stage, to clarify terms a little. Some would argue that the term 'model' has a very wide connotation, and represents a description of anything which abstracts reality in some way. But most would say that iconic models are those which represent the real world with *only* a change in scale (e.g. an aerial photograph); analogue models are those in which one property is represented by another (e.g. contours showing relief on a 2-D map); symbolic models are those in which symbols represent objects or properties (e.g., signs on a map, a mathematical equation to describe a relationship). This wider usage of the term helps in the reconstruction of geographical methodology since it illuminates the nature of much 'traditional' work done in geography classrooms and relates it to newer developments. It also helps to identify the nature of physical representations of reality ('hardware' models) which have been part of the geographer's stock in trade for many years.

But beyond this, the term 'model' is in current usage more specifically to describe a range of simulation techniques in which 'dynamic representations employ substitute elements to replace real or hypothetical components'.

J. L. Taylor[1] recently classified these according to their degree of abstraction from the real life operation and suggested the following:

Reality←---				---›Increasing abstraction
Case study or	In-basket	Incident process	Role playing	Gaming simulation	Machine or computer simulation
	in-tray method				simulation

Case studies simply involve the detailed description or history of selected problem situations—geographers know this situation better as the 'sample study'. The *in-basket* method is a technique in which one person is asked to consider and act on a set of issues— say those that a morning's post might bring in. The *incident process* is a variant on the case study, in which the participants seek for additional information as they work.

Role play is more elaborate than the first three mentioned because it relies on spontaneous mock performances from a group of participants, placed in a hypothetical situation. It may also form part of *gaming simulation*, in which human decision-makers are confronted with varying situations to test their reactions to them.

Computer simulation, on the other hand, is one in which all data and decisions are embedded in a machine. Once a programme has been made, human participation is limited.

Case studies and role play are of interest to geographers in many ways, but this particular text is most concerned with the family of models which Taylor identifies as 'gaming simulations'. The use of the term may perhaps be a little ambiguous in those simulation models where competitive elements are not stressed. The American High School Project's 'Portsville' and 'Metfab' simulations[2] are examples of this; as are the town building simulations developed for English classroom conditions by J. A. Everson and B. P. FitzGerald.[3] But many 'gaming simulations' do have a competing element, and it is these that are now considered in the pages that follow.

The history of operational games

In recent years, the operational game has made its way with considerable success in business and management. American techniques of executive training have incorporated sophisticated role play situations and gaming simulations into their programmes and the idea has made its way with notable success into other fields.[4]

This tradition goes back to the development of conflict and war games from at least the nineteenth, if not the eighteenth, century used to train officers in strategy and tactics.[5] A devoted band of enthusiasts today make the art of 'war gaming' a colourful and less sober affair by the re-enactment of famous battles with lead models at various conclaves which they hold.

Another strand of gaming derives from the philosophical trend in the amusements provided for Victorian children. The jigsaw puzzle was, in fact, invented in the eighteenth century by a firm of cartographers, and originally called the 'dissected map', an indication of its first subject matter. Jane Austen was aware of the use of this rudimentary geographical game; in *Mansfield Park*, Fanny is criticized for her inability to 'put the map of Europe together' when playing with her cousins.

A later game, dating from 1843[6], had a more developed process situation, with four contestants—a pedestrian, a steam-boat, a train and a futuristic 'aeroplane'—involved in an 'Eccentric Excursion to the Chinese Empire', in a colourful if geographically dubious expedition.

This type of game was the popular forerunner of parlour games like 'Ludo' and 'Snakes and Ladders' which many modern families enjoy. The enjoyment and excitement which these generate at each playing, especially with young children, are a testimony to the soundness of their construction. It should not seem strange, therefore, if games are used as an educational tool in the classroom. Changing values and techniques in the primary school have prepared the way for this; it has been demonstrated that quiet classes are not necessarily learning at an advanced rate, nor those who sit in orderly rows. The liberation of children into 'activity methods' has already had rewarding developments at the lower age ranges.

But in the secondary school methods are not yet as informal. The suggestion that games might be used in a geography lesson has on occasion been greeted, in my experience, by a frigidity bordering on pity, if not contempt. Even at university level, half-apologetic disclaimers have to preface discussion of game-theory. 'The Theory of Games' despite its immediate connotation of amusements of a frivolous kind, is an imposing structure, dealing with. . . .'[7] Perhaps it is not necessary to apologize for frivolity in the classroom. Lack of humour and lack of enjoyment are no passport to learning; and as McLuhan has reminded us, why should we think that our good students are our *serious* students?

In other subjects, besides geography, the possible use of gaming-simulations has lately been investigated with some seriousness. Several conferences and books have developed the idea on a variety of fronts.

Some effects of games

Children may well become highly motivated to play a classroom game, and to enjoy it at the same time. Such enjoyment is no bar to learning; on the contrary it is likely to increase not only the desire to participate, but the readiness to receive ideas and information which may emanate from the simulation itself.

Children often learn best from their contemporaries—a fact that we might investigate more than we do. One teacher friend of mine, writing to me about the success of a simulation with which he had experimented, amplified the point graphically:

At one time talking during written work annoyed me greatly. But after I gave 4b and 4c written question sheets based on a duplicated information sheet, I really began to see the value of 'private discussion'. I can't tell you the satisfaction it gave me to overhear things like this (an actual example):

JOHN: You done 2A yet?

FRED: Yes, it's easy.

JOHN: I can't see it.

FRED *(pointing to data sheet)*: Look, it says production will be speeded up by the introduction of new bottling machinery and the delivery of grapes by helicopter from the vineyard.

JOHN: How's that show more mechanization?

FRED: Bottling machines and helicopters. They're machines, aren't they? Mechanization means using machines, don't it?

JOHN: Oh, yes.

It seems to me that games develop this sort of conversation and justify themselves on these grounds alone.

Some of the elements of game playing are of considerable importance in developing relationships and personalities in the classroom. Games in which groups of pupils need to discuss and make decisions are practical and simple examples of the kind of democratic procedures which Dewey thought important. The need for cooperation among pupils is given a chance to make itself clear at an easy and practical level. The team which finds itself constantly haggling over internal decisions will probably penalize itself in its external relations; but equally, the price of false unity may be capitulation to the rashest or most forceful in the group.

Similarly the opportunities for extended role-play in the games allow the development of imagination within the individual. The kind of work done in this geographical context emphasizes the importance of drama not so much as a school subject but as a method of work.

There are cautions to add. An unvaried diet of games or an over-extension of them beyond a pupil's span of concentration is as unwise as a constant diet of note-taking or textbook reading. The games in themselves do not offer a panacea in dissident geography classes that allows sweetness and light to reign. They are simply an extra technique which may help on occasion in developing modern geographical teaching.

There are dangers in the 'competitiveness' of the games. Though most children beyond the age of nine are able to take 'losing' in their stride, there may be some for whom the experience is too intensely bound up with their own personal inadequacy for it to be seen as 'fun'. For them the game and its purpose have to be carefully interpreted by the teacher; it may be more useful for them to operate the game as a personal, rather than a group, activity. The challenge to the introvert child may be in playing and understanding the game to his own satisfaction, rather than competing within it. Group participation alleviates this personal problem to some extent.

R. E. Kasperson has also importantly noted possible limitations.[8] He points out the difficulty of measuring learning in these situations, the possibility of transfer *not* taking place, and the danger that winning rather than learning may become the emphasis. It would be unwise not to consider these points when game enthusiasm seems to be full of promise.[9]

Description of the games that follow

The six games which follow in Chapters 4—9 (some examples of games) are designed for use in the classroom. They can be played either by individuals or, perhaps more realistically, by groups of individuals operating as teams. They are meant to illustrate fundamental geographical processes—the problems of developing a country, of building railways, of exploration for minerals, and so on. Though given one specific context they are meant to be transferable (in essence) to any other. The transfer can be made by redesigning the base map of the game, or by applying the principles of the game to another situation after it has been played.

The games are meant to combine acceptable educational techniques with relevant subject matter in an atmosphere of informality and enjoyment. They are, of course, part of the model theory already referred to. They simplify reality in their assumptions and rules and bring into play what the inventor of the game considers to be the crucial factors in a particular situation. But the players of the game may not necessarily be aware of these conscious hypotheses as they play the game, though they enter into the area of real consideration and decision by the act of playing.

In this there is the real chance to accelerate the learning process since (in Piaget's terms) the game is a 'concrete operation' which can deal with a good deal of advanced material long before that material can be consciously absorbed at the level of formal concepts.

Thus the game can be played at two (if not three) levels: that of enjoyment for its own sake; that of conscious understanding of hypotheses involved; and that of conscious attempt to improve the simplifications made—a redesigning of the game to approximate more closely and bridge with reality.

The simplifications have their value. They are an aid to understanding the basic dynamics of processes uncluttered by the 'background noise' of transitory or irrelevant material. The *kinds* of decisions which railway builders face (see Chapter 7) do not differ in essence—they are concerned mainly with the problems of economics on the one hand, and with physical environment on the other; it is the environmental context which varies from place to place, as a kind of cloak to the basic considerations.

Is it more important to observe each of the differing cloaks? Or to discover the considerations which lie beneath? This is perhaps at the heart of the discussion about geographical objectives in modern education. In the past, it seems to me, our teaching has sometimes tended to be a dazzling fashion parade of all the cloaks. They have been paraded before the eye of the awestruck viewer for an instant, but then have moved on before the teacher has really had time to reveal what lies beneath them.

The use of games to teach some fundamental subject structure at a highly motivated level is therefore a prime objective. It is also an attempt to improve the attitude towards problem solving in an elementary way.

> Mastery of the fundamental ideas of a field involves not only the grasping of general principles, but also the development of an attitude towards learning and enquiry, towards guessing and hunches, towards the possibility of solving problems on one's own. . . . To instil such attitudes by teaching requires something more than the mere presentation of fundamental ideas.[10]

The understanding of fundamentals is therefore one aim of this approach. A better attitude towards motivation in the classroom is another. A more active development of cooperative relationships is a third. These fundamentals may also become more comprehensible, more easily remembered, and more easily transferred to other problems. Comprehension of them, at whatever level, is the vital bridge that can keep the two 'poles' of the subject (at primary school and at university research level) connected with each other. Otherwise geography at one end of its scale may rapidly become unrecognisable from the other.

The six games which follow do not differ from their commercial counterparts in method but in aim. Their aim is not to be the 'most exciting game possible', but the 'most realistic game possible'. Thus some possible fantasy elements have been excluded from the games, despite the fact that they might have helped to 'improve' the game. On the whole pupil response to the games has suggested that there is in any case no pressing need for artificial elements.

The operation of the game has been devised so that geographical understandings (amongst others) become important to the successful playing of it; the teacher who uses these games has therefore a vital part to play in exploiting and reinforcing these understandings in later teaching.

Each of the games is based on a real geographical context—a springboard from which the game was devised. The first game, SHOPPING, for example, is based on the layout and character of a real shopping centre. In playing it pupils face, in a simplified version, the problems that face shoppers using that particular centre. These problems are similar to those in many other shopping centres, and if a teacher cares to devise a base-map of his own shopping centre, the new game can be played within the broad general rules of the original. Similarly, though RAILWAY PIONEERS is based on the real context of the building of transcontinental railroads in the USA, it could be adapted and transferred to Britain or Africa.

The aims of the games are set out explicitly in the list of instructions which go with each, but it may be worth noting here some broad areas of concern which can be identified.

1 The games deal with important basic concepts and ideas within the framework of space, e.g.:

the idea of 'friction of distance' in EXPORT DRIVE;

the idea of 'the connectivity of networks' in BUS SERVICE;

the idea of searching an area efficiently in NORTH SEA GAS and DEVELOPMENT.

2 The games attempt to deal with the understanding of important decision-making processes that are relevant to human geography. Pupils play the roles of people who take such decisions, and are able to see through to the basic problems because of the way in which the games deal with the real world in a simplified way, e.g.:

the decisions made by railway builders in RAILWAY PIONEERS;

the decisions made by exporters in EXPORT DRIVE;

the decisions made by mineral exploiting firms in NORTH SEA GAS.

3 Some of the games use large regions of the earth's surface as their base. Factual material concerned with world geography will be conveyed through the playing of the games, and indeed will be learnt in order to play the game successfully, e.g.:

world routeways and the position of some countries in EXPORT
DRIVE;

Some physical and human geography of the USA in RAILWAY
PIONEERS;

The offshore geography of North-West Europe in NORTH SEA GAS.

4 The games involve the structuring of geographical material
into meaningful categories and classifications. By using some of
these simple arrangements of geographical material, some
insight is thrown on the importance of structure; factual
material is not seen merely as a confetti, e.g.:

the idea of urban hierarchies is used in BUS SERVICE;

the idea of shopping 'quarters' is used in SHOPPING.

As in many commercial games, chance factors—the throw of a
dice or the turn of a card—have their place as a characteristic in
this set. This is not a concession to fantasy in the interests of
excitement; quite the reverse.

The function of chance in random selection was, as D. R.
Stoddart has recently pointed out, one of the major omissions in
what geographers chose to adopt from the fruits of Darwinian
theory. Only recently have belated attempts been made to
recognize it and restore it as a realistic factor in various situations.

Because of the omission, geography went through an era of
determinism, or neodeterminism (in which crops grew in certain
places *because* the climate was right, for instance, and apparently
without regard to the needs, aspirations, or choice of the
inhabitants). This brought disrepute to the subject through its
patently obvious over simplification of environmental situations.
Of recent years, the crudity of the concept was ironed out, but
a deterministic 'hangover' still pervades some classroom geography,
if only as a kind of self-justification for the subject.

The need for more precise work in the social sciences caused an
early use of stochastic techniques there, and following in these
footsteps geography has also recently accepted the use of these
techniques as a step forward in helping to analyse human
geographic problems.[11] The acceptance of chance as a factor in
decision-making has not only made it more possible properly to
evaluate existing problems in precise terms; it has also made the
whole study of them more realistic.

Thus, the use of dice and chance-factor cards in the games should not, I hope, develop doubts about their realism—rather the reverse. The controlled operation of such a factor is a more correct simulation of human processes than one which has invariable predetermined reactions to specific situations.

Each of the games also involves some very simple computation in arithmetic processes—the keeping of a monthly balance sheet, or the calculation of an annual expenditure, for example. The maths in these is nothing beyond what a modern primary school child would expect to do at the age of 9 or 10.

In this sense, these game models are almost non-quantitative. Anyone with a mental block about simple arithmetic may be urged to conquer this in order to cope with the simple digit sums that each game may require. The simplicity of the maths may also serve to emphasize that models are not simply for the 'high-flyers' and that the quantitative revolution is not necessarily a demanding corollary of their use.

The games in the following chapters range in complexity: SHOPPING has been played in a primary school with some success, as has BUS SERVICE and RAILWAY PIONEERS (the latter two in simplified versions from the rules printed in Chapters 5 and 7). But as written here they are designed for secondary school use. At least the first five can be used in the lower part of the school; only EXPORT DRIVE perhaps needs to be confined to fifth and or sixth formers.

The rules as suggested are not unchangeable. Notes at the side of some of them indicate where they can be omitted if required. The younger the children with whom the games are played the more complexities it may be wise to omit. In this case the game becomes simpler and easier to play without losing its fundamental character; but it becomes a more approximate model of the reality it tries to represent.

As the ability of the children using the games increases, so should the complexity of the game. With some experience, therefore, pupils can handle models which are more refined than those with which they began.

Younger children (perhaps mainly those between the ages of nine and thirteen) will enjoy playing the game for its own sake.

The attitude of older pupils will be more sophisticated. They will probably become involved in the game in the same way as younger pupils, but also be able to adopt an adult stance towards game playing and view their own actions and participation in an objective way.

Older children, for example, will be able to assume the 'roles' of entrepreneurs or explorers more effectively through some previous knowledge of the way that these men are likely to act; younger pupils will simply obey their own instincts and make their decisions in a subjective personalized way.

Older pupils will also be able to contribute more to actually devising or improving a game, or to simplifying it for use with younger pupils. In improving existing ideas (such as those in the following chapters) they may be able to devise further sets of rules which can be incorporated to bring the model even closer to the reality it seeks to represent. In doing this, it is perhaps worth noting that the frequent pitfall is to reduce the practical possibilities of 'playability' in schools to zero through overzealous sophistication. But in the act of attempting improvements, pupils can often help themselves to analyse reality more effectively as they search for a simulation of a process, an effect, or a set of actions.

In simplifying a game, older pupils will reveal whether they fully understand the basic factors involved. (They may have been playing the game with limited success without fully appreciating what they were doing.) One fully appreciates that the teacher does not always have much time to spend on complicated organization, preparation or explanation of projects such as these.

I have seen a sixth-form group of geographers lighten a teacher's load very effectively, by playing an operational game and then simplifying it in readiness for its use with a first-form. The older pupils themselves went into the lower form, explained the purposes and basic points of the game, and themselves 'managed' its operation, whilst giving useful advice. They said afterwards that they 'thought they had probably learnt more the second time than the first'.

Older pupils may also be motivated to construct games of their own, using their own geographical knowledge, and choosing material relevant to their own particular syllabus. Indeed, it would be sad if they did not. The intention in publishing this particular set of games is not to offer them as defined and rounded end-products in their own right; they are in print simply to show the kind of game that can be developed, and to encourage others to take them and adapt or revise them for their own purpose. Chapter 11 in this book offers some help to those who wish to develop games of their own.

If older pupils attempt the task of devising their own material the construction process itself and its relation to the interplay of factors will clearly reveal whether they appreciate the real dynamics of the situation under consideration.

Practical points

Each individual teacher will know best how to use these games in his own particular situation; a couple of representative examples of the use of operational games in a syllabus is described in Chapter 10 for those who wish to read of actual experiment.

The games can be used by either four or five players individually (a group within a class or a small tutorial group in the sixth form, perhaps) or alternatively by four or five groups within a class. The groups can consist of anything up to six or seven players each, since there can be a delineation of specific roles in several of the games, and a corporate decision-making process in all of them. In the games where groups are required to simulate the decisions of competing companies or organizations, the large size of the group is an advantage. The interplay within the internal group arrangement is itself a useful place in which to see some of the factors being given serious discussion.

In most cases, if the game is played as a class activity, it is wise for the teacher to act as 'umpire' or 'administrator', though this role can be assigned to a member of the class, or to an older pupil. Some games do not require this function (e.g. BUS SERVICE and SHOPPING). If the game is played with small groups, one of the group may act as 'umpire'.

With four or five individual players the games will tend to be played more quickly than those in which groups participate.

Some of the games may be played through within 30-45 minutes in this fashion, and can be used in a single period. They can be given to a group of more (or less) able pupils during a period when a variety of tasks are being performed. They offer a little relief to the 'find out what you can about x' approach, and may certainly be played more than once by the same group of pupils in most cases.

If the games are played with larger groups they will take longer; the individual decision processes themselves being subject to discussion and majority voting. In RAILWAY PIONEERS for instance, the company may be set up in quite elaborate form (see Chapter 10). There may be proper board meetings, appointment to posts such as secretary, surveyor, and engineer, and agendas for board meetings. The minutes of such meetings and the balance sheets kept by each treasurer may be required for homework for each group, and the game may go on over a period of several days. In this case, decisions can be pondered and discussed, and the game may be played in short spasms at the beginning or end of class periods, or even out of school time altogether. There is no reason why other activity should not be interspersed between stages of development of a game.

In any case, it is advisable for the teacher to explain the idea of the game briefly to the participants before they begin, in order to supplement written instructions. In this way, the basic purpose of the game can be stressed, and any major questions or misunderstandings dealt with.

Having explained the game, and then allowed a group of pupils to play it, a teacher may find that some members of the group immediately want to repeat the exercise. This seems to happen if a group gains some expertise during the game and wishes to profit from its new proficiency, or if it feels hard done by through chance factors, or if it feels that it now understands the purposes and strategies involved. I have found that playing a second time is often a beneficial experience for those who wish to take part. By this time a high motivation to repeat the sequences of the game allows them often to penetrate deep into the interplay of factors which the game seeks to ulluminate. It has not been unusual to see, at this stage, geography being done 'out-of-school' as groups come in to operate the model in lunch hours, or after school. Such motivation is relatively rare in our more orthodox techniques! But for some, one 'taste' of the game will be enough. Though they may wish to watch or help others on a second run, it cannot be expected that games will gain a universally enthusiastic response.

One of the obstacles to game-playing is obviously the provision of materials. It has not been practical in this volume to provide actual equipment for the games, but only to explain and illustrate what is needed. The provision of the actual material would have been impracticable, and perhaps undesirable. Practical suggestions are made in an appendix (p. 117). However, senior pupils seem to enjoy the construction of game kits as much as the playing of the games themselves, and the process is not a time-consuming one.

In most cases a large base-map or plan, suitable for 'public' exhibition is needed. It is perhaps better if it is exhibited in a vertical position so that pupils can sit in their usual positions and observe the game's progress. A base-map of this kind will need counters or symbols which adhere in this position. Some suggestions are made in the appendix concerning the kinds that may be used.

The vertical position is, in my experience, to be preferred to the plain sheet or card or paper placed horizontally on a desk, since this causes viewing problems for large groups (although possible for small groups). An additional hazard, however, is that counters placed flat on such a map may become disarranged if they are not of the adhesive type.

The large map may be supplemented for large classes by duplicated foolscap-size maps. If these are provided, one for each group, they can cut down class movement, and enable each board or company to plan its own strategies in more seclusion. The large map, handled by the umpire, can then be used as a 'master map', and plans for future exploration or development can be plotted or hypothesized easily if the groups themselves have a supply of maps which have previously been prepared.

GAMES IN GEOGRAPHY

References

1 TAYLOR, J. L. *Some pedagogical aspects of a land use planning approach to urban system simulation*, University of Bristol, Dept. of Geography, 1968.

2 GRAVES, N. J. (see ch. 2, note 12).

3 Described briefly in WALFORD, R. 'Decision-making', *The Indian Geographical Teacher*, Vol. IV, Pt. I, Dec. 1968, pp. 17–26.
Everson and Fitzgerald have a book *Inside the City*, in the press (Longman), which will give further detail on the operation of these models.

4 See CLUSHEN, W. E., ed., *Operational research for management*, vol. 2. Johns Hopkins Univ. Press, 1956, chapter by MCCLOSKEY, J. F. and MCCOPPINGER, 'Operational gaming in industry'.

5 YOUNG, J. P. *A brief history of war-gaming*, Johns Hopkins University, Department of Operational Research, 1956.

6 In the collection of MUIR, P. H.

7 GOULD, P. R. 'Man against his environment; a game theoretic framework', *Annals of the Association of American Geographers*, September, 1963, pp. 290–7.

8 KASPERSON, R. E. 'Games as educational media', *Journal of Geography*, Oct. 1968, Vol. LXVII, No. 7, pp. 409–422.

9 See also reservations noted by TAYLOR, J. L. and CARTER, K. R., 'Instructional simulation of urban development—a preliminary report', *Journal of the Town Planning Institute*, 1968, p. 445. These refer to complex games played with university students.

10 BRUNER, J. S. *The process of education*, Random House, 1960, p. 20.

11 A good introductory discussion is to be found in *Frontiers in Geographical Teaching*, ed. R J. CHORLEY and P. HAGGETT, Methuen 1965, pp. 110–115.

4

The Shopping Game

1. The operational game SHOPPING is one in which players or groups of players assume the roles of shoppers in a town centre. They are given a list of shopping tasks to carry out and seek to do this in the shortest possible time.
2. The aims of the game include:
(*a*) to understand something of the internal structure of shopping areas, i.e., that different 'quarters' develop, specializing in particular retail goods; that shops often cluster around a central node but also stretch away from that node in sectors etc.
(*b*) to develop an insight into 'route efficiency'.

Basic ideas

THE REALITY

1 Retail service areas in the centres of towns are usually clustered around a central point, the 'down-town' district or the 'central area'.

THE GAME

The game is played on a map on which is drawn any real or imaginary shopping area. Shops (and their type) are also indicated on the sheet, and the pavement areas are shown. (See Fig. 1 for a sample plan.)

The pavement areas are divided into grid squares. It is assumed that it takes five seconds to pass through every grid square.

Roads may be crossed by zebra crossings—these take fifteen seconds to cross.

2 Shops can be categorized into types, e.g. food shops, clothing shops, other retail goods shops, shops offering a service, public buildings, etc.

The shops on the map are divided into four different types and may be coloured accordingly. Public buildings (Library, Post Office, and banks), clothing shops, food shops, other retail goods shops and shops offering a service.

3 Many people go to their local town shopping centre by public transport. Bus stops are conveniently sited to the central area.

Eight bus stops are marked on the map; four are for Route X, two on each side of the road, four are for Route Y: likewise.

4 Though adults may wish to be reflective in their shopping, i.e., stop and consider quality, talk to friends, etc., younger people often wish to achieve the task as quickly as possible.

The aim of the game is to begin from a bus stop, visit a certain number of shops to purchase goods, and then to return to a bus stop on the *other* side of the road as quickly as possible.

5 It is possible either to buy goods from a number of limited-commodity stores; or perhaps to buy a greater number from a supermarket or a chain store.
If the latter course is taken, it is *also* possible to be

Players are given shopping lists. The lists identify in which shops items can be bought.
It is possible to buy *all* food items in the supermarket; and *all* retail goods items in the chain store; but there is

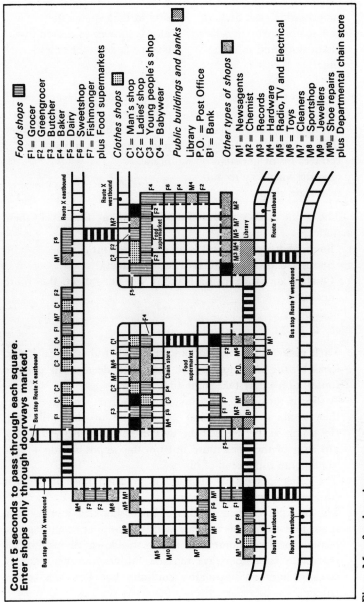

Fig 1. Map for shopping game

caught in a long queue at the supermarket cash desk, or to lose one's way in a large chain store and so waste time.

a fifty-fifty chance of being caught by a queue there. (A dice is thrown to decide this chance factor—see rules.)

6 A person who plans his or her journey sensibly will do the shopping more quickly than one who continually retraces his or her steps or goes the long way round to certain shops.

The aim of the player is to complete his or her shopping list in the shortest possible time and return to a bus stop for home.

The game makes certain simplifications at this level; these include:
1 Buying each product takes an equal amount of time.
2 Shops are always in stock of each product.
3 Quality or price considerations do not affect the route decisions.

How to play

1 Each player draws a 'shopping list' and studies it. Each also looks at the shopping centre map and decides on possible routes. (This may take some time.)
2 A dice is thrown to see who should go first. The first player then nominates *from which bus stop* he or she will start. (In doing so, he needs to be aware that he must return to a bus stop of the *same* route but on the opposite side of the road at the conclusion of his task.)
3 The other players do likewise in rotation and place their counters at their starting points.
4 The first player then nominates his or her first journey to a shop, and traces this journey with the counter, counting five seconds for each square, and noting this down. Shops must be entered *through* doorways marked. Moves can be made in straight lines *or* diagonally, counting five for each square entered.

5 Each of the other players does likewise for his or her own first journey.

6 The first player then nominates his or her second journey, and traces the route with the counter, and so on.

If players choose to take a chance by going to the supermarket or chain store in order to try to buy more than one product at a time, they must throw a dice. If the dice comes down 1, 3, or 5 they have judged correctly and speeded up their shopping; if the dice comes down 2, 4 or 6 they have run into queue problems and must add on a *delay* figure of 60 seconds to their score.

7 When the purchases have been completed the player returns to a bus stop on the opposite side of the road. It need *not* be the stop which is directly opposite to the one where he got off, but he must return on the same *route*, i.e. he could catch a bus somewhere else in the centre.

8 It is interesting to see who has shopped fastest. It is also often useful to change lists round and see if other people can beat the original time in subsequent rounds.

Equipment

1 Map of shopping centre, with shops and pavements adequately marked. This map can either be made from a known or local area (perhaps as adjunct to local study work) or else the accompanying map can be used.

The map can either be enlarged so that all players move their markers on one board; or kept at foolscap sheet size, so that each player enters his own moves on his own sheet.

2 Dice and shaker; markers to move on maps.

3 Set of shopping lists. These should have six items marked on them and include visits to each of four types of shops. For example:

LIST I

Potatoes	Return library books
(Greengrocer)	(Library)
Biscuits	Buy gramophone record
(Grocer)	(Record shop)
Pint of milk	Buy handkerchiefs
(Dairy)	(Men's shop)

LIST 2

Book of postage stamps	Buy new school tie
(Post Office)	(Men's shop)
Pair of shoelaces	Fresh fish
(Shoe repairer)	(Fishmonger)
Take brooch for mending	Sausages
(Jeweller)	(Butcher)

LIST 1 has three food items, one public building item, one retail goods item, and one clothing item. It might be worthwhile for the player to try the supermarket with three items in the food category on this list.

LIST 2 has only two food items, but also two retail goods items, one public building item, and one clothing item. A different strategy is probably called for.

Players need to note that there is often more than one shop selling a particular item in the shopping centre.

4 Simple time sheets in which to add up the time of journeys taken. For older pupils this can safely be done on scrap paper and then checked.

Some possible complexities

SHOPPING is basically a fairly simple game. If desired, certain complexities can be introduced in order to bring the game closer to reality.

1 The shopping list can be lengthened to eight or ten items, and certain 'heavy' items starred. For instance potatoes or library books might be thus marked. These items would have to be put into the mythical shopping bag *first*. They would need to be bought *before* the biscuits or the cream cakes or the eggs.

2 The possibility of items being out of stock can be introduced. A dice can be thrown for each purchase. Throwing a 6 would mean 'out of stock' and the shopper would have to go elsewhere.

3 'Conversation' stops can be built in as chance factors by dice throw (e.g. each player has one throw at the end of each round and if a 6 is thrown, he or she is delayed two minutes by conversation).

4 There is *another* geographical exercise involved in analysing which shops are visited most frequently, and why.

5

The Bus Service Game

1 The operational game BUS SERVICE is one in which players or groups of players assume the role of bus operators. They assess the potential of different routes and apply for licences to run them. Their aim is to build a coherent bus network with maximum passenger carriage.

2 The aims of the game include:

(*a*) to understand the idea of connected and disconnected network systems and to see how this abstract idea affects decisions in the real environment;

(*b*) to understand the basic principles of urban hierarchies and the relation of trade and traffic to this.

Basic ideas

THE REALITY

1 Bus operators usually operate between areas of settlement by using the A or B roads.

THE GAME

The game is played on a map which represents the major roads of the Isle of Wight (see Fig. 2). A roads are marked in thick lines and B roads in thin lines.

2 In the present day, most competition between bus companies is not in running over the same routes; it is

Players in the game need to consider the map and then decide which licences they most wish to apply for.

in the application for licences for an exclusive route.

One player (or a separate observer) acts as Minister of Transport and deals with the issuing and returning of licence cards.

3 Settlements are of different sizes.

The towns on the map are graded into four different sizes in order to emphasise the urban hierarchy. These are GRADE A settlements population 10,000–20,000 (marked in capitals). **Grade B** settlements population 5,000–10,000 (marked in bold type). **Grade C** settlements population 2,000–5,000 (marked in lower case). *Grade D* settlements population 1,000–2,000 (marked in italic)

4 The largest number of passengers per day is likely to be carried on routes between the largest towns. Generation of traffic is related to the size of settlement.

There are licence cards for route between adjoining pairs of places on the map. These carry information about the average number of passengers on that route per journey. (All bus operators are presumed to operate the *same* statutory number of journeys.) The number of passengers ranges from twenty (between two A settlements) to two (between two D settlements).

5 Companies compete for licences but are not always successful in their appli-

Players apply for licences of their choice one by one. But they also throw a dice to

cations. Sometimes, perhaps because of inefficiency or public complaint, licences are withdrawn.

6 Efficient networks are created if companies obtain licences between adjoining towns so that buses can be used in an economic way on more than one route.

7 Sometimes occasional traffic to certain places of leisure or entertainment swells the passenger load of a bus company.

include a chance factor in the application (see 'How to play' for more details).

Companies seek to obtain licences of routes that adjoin each other. If they achieve two adjoining licences they can *double* the scores on the licence cards; if they achieve three or more adjoining licences, they can *triple* the scores on the cards; if they achieve a completed circular network they can *quadruple* the scores on the cards. Thus an efficient network organized among small towns may produce more traffic than an erratic network linked to big ones.

After all the licences have been distributed, each player also draws two Occasional Traffic cards from the pack. If these apply to routes which he owns he may add any bonus numbers to his own passenger total.

How to play

1 The map and the licence cards are laid out. The players throw a dice and the person who throws highest has the first chance of licence application. The others follow in clockwise rotation in turn.
2 The first person chooses a licence for which he wishes to apply

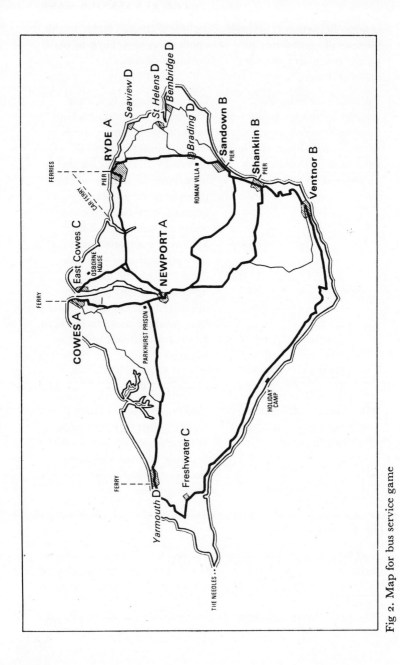

Fig 2. Map for bus service game

and nominates it. He throws a dice. If he throws 1, 2, 3 or 4 the licence is granted, and marked on the map. If he throws 5 or 6 it is *not* granted. If he throws 6 when in possession of a licence or a number of licences he is required to *return* one to the licensing authority. (The one he returns can be of his own choice.)

3 The second person then also nominates a licence and throws a dice as in 2 above.

4 The other players do likewise.

5 This continues until all licences are finally in the possession of the players. Players will obviously try to build up networks of routes and achieve licences that run between the largest centres of population if they can.

6 Next each player draws the Occasional Traffic cards from the pack.

7 Then totals of passengers are counted. Licence cards may be used as doubles, or triples or quadruples but no one card may be used in more than one set. The figures for occasional traffic are added in as bonus *if* the player owns the routes that are given bonuses from the cards he draws—*not* otherwise.
(Indications of the occasional traffic potential can be seen from the map; holiday camp, yachting, ferry service, piers, etc.)

8 The person with the highest passenger total is the most successful bus operator.

Strategy

1 To win, it is not necessary always to have the licences between the largest towns. A good network elsewhere can produce results.

2 The Occasional Traffic cards may sometimes be the deciding factor at the end of the game, and the occasional traffic indications on the map may give some routes more attraction than their termini suggest at first glance.

3 The cleverly placed application (if successful) can disturb the network ambitions of rivals.

Though chance will have determined some of the results, the

person with the best eye for networks and hierarchies often becomes the winner.

Equipment

1 Map of Isle of Wight (with major towns and roads marked). This can be (*a*) one large one; (*b*) set of personal ones; (*c*) both, see Fig. 2.
2 Licence cards (see Fig. 3).
3 Occasional Traffic cards (Fig. 4).
4 Markers to mark routes for owners, on the large map; dice and shaker.

TRAFFIC AMOUNTS BETWEEN SETTLEMENTS

A settlements—10,000-20,000 population
B settlements—5,000-10,000 population
C settlements—2,000-5,000 population
D settlements—1,000-2,000 population
Traffic between
A and A 20
A and B 16
A and C 12
A and D 8
B and B 14
B and C 10
B and D 6
C and C 8
C and D 4
D and D 2

e.g. Between Newport and Ryde, two 'A' grade settlements, twenty passengers can be assumed per journey.

SUGGESTIONS FOR OCCASIONAL TRAFFIC CARDS

| 1 Visitors flock to the Shanklin Pier Show | *Bonus* 20 Newport–Shanklin route 20 Sandown–Shanklin route 20 Ventnor–Shanklin route (if owned) |

2 Heavy traffic from the main-
land; London holidaymakers
arrive by ferry in large numbers

25 Ryde–Brading
   Brading–Sandown
20 Ryde–Seaview
15 Seaview–St Helens
10 St Helens–Bembridge
20 Cowes–Newport
30 Cowes–Yarmouth
40 Freshwater–Ventnor

3 Brightstone holiday campers visit
Ventnor for the day, en masse
4 New exhibits at Osborne House
attract visitors
5 Public meeting in Newport
draws large crowds to protest
against further prisons being
built on the island

25 Newport–East Cowes
25 Ryde–East Cowes
20 Yarmouth–Newport
15 Cowes–Newport
15 Ventnor–Newport
15 Sandown–Newport
15 Shanklin–Newport
20 Ryde–Newport

6 Bembridge regatta day

40 Sandown–Bembridge
20 St Helens–Bembridge
10 Seaview–St Helens
40 Yarmouth–Newport
30 Yarmouth–Cowes

7 A Southampton-Cowes ferry
has engine trouble; traffic
diverted to the Yarmouth-
Lymington (Hants) link tem-
porarily
8 A class at Newport Comprehensive School are taken on a
geography field trip by bus to
look at stack formations and
reverse faulting at the west end
of the island

40 Newport–Yarmouth
40 Yarmouth–Freshwater

9 Sandown Water Sports

20 Newport–Sandown
40 Sandown–Shanklin
25 Brading–Sandown
20 Bembridge–Sandown
20 Seaview–St Helens
20 St Helens–Brading

10 A group of American tourists
decides to visit the Roman villa
at Brading by bus.

11 Sailing races at Yarmouth

12 Cowes Regatta week

20 Newport–Yarmouth

40 Newport–Cowes

25 Yarmouth–Cowes

Fig 3. Specimen licence card   Fig 4. Specimen Occasional Traffic card

**Licence to operate
a service between**

NEWPORT
(Grade A settlement)
and
SHANKLIN
(Grade B settlement)

average number of
passengers: 16

Occasional Traffic Card

Visitors flock to
SHANKLIN PIER SHOW

Bonus:

20 NEWPORT-SHANKLIN

20 SANDOWN-SHANKLIN

20 VENTNOR-SHANKLIN
                    if owned

Make licence cards for:

Newport-Ryde
Newport-Cowes
Newport-Shanklin
Newport-Sandown
Newport-Ventnor
Newport-Yarmouth
Newport-East Cowes
Ryde-East Cowes
Sandown-Shanklin
Shanklin-Ventnor

Freshwater-Ventnor
Cowes-Yarmouth
Yarmouth-Freshwater
Sandown-Brading
Sandown-Bembridge
Ryde-Brading
Ryde-Seaview
Seaview-St Helens
St Helens-Bembridge
St Helens-Brading

6

The North Sea Gas Game

1 The operational game NORTH SEA GAS is one in which players or groups of players assume the role of companies searching for gas in the North Sea. They consider what kind of drilling rig to use, where to search for gas, and then whether or not to exploit it.

2 The aims of the game include:

(*a*) to understand the problems facing companies searching for fuels or minerals.

(*b*) to learn specifically about the problems of North Sea gas exploitation, and the factors associated with this as an economic operation.

(*c*) to gain experience in the technique of search and to find out how this can be done most efficiently.

Basic ideas

THE REALITY

1 Late in 1964 the first drilling rig began operations in the North Sea. Companies who began a search for gas and oil had little information on which to work, and no great source of equipment to hand.

THE GAME

The game is played on a board which represents the U.K. section of the North Sea, assigned under international agreement (Fig. 5) (1964).

2 The U.K. government marked out the area in concession squares and invited companies to bid for these areas.

3 Companies began drilling with an assortment of drilling equipment. Some used *self-elevating platforms* which could only operate in relatively shallow water (up to one hundred and fifty feet deep).
Others used *floating drill ships* brought from elsewhere.
Still others sought to hire or buy the larger *semi-submersible* platforms.

4 Some of this equipment was hired for three-month periods. Other companies bought outright the rigs they were using.

5 Companies had little evidence on which to base decisions for drilling— though it was known that gas had been found in commercial quantities in Holland, and there were small gas fields in Britain. The size of gas domes was also roughly known. Other

The map is marked out in concession squares, approximating to the divisions made under the agreement.

Companies have the choice of using one of three kinds of rig for drilling; (*a*) self-elevating platforms; (*b*) floating drill ships; (*c*) semi-submersible platforms. Self-elevating platforms only operate in shallow areas and do not drill if the wind is above Force 7.

Floating drill ships operate in all areas but do not drill if the wind is above Force 5.

Semi-submersible platforms operate in all areas but do not drill if the wind is above Force 8.

Companies have the choice of hiring or buying their rigs. To buy outright is cheaper but less flexible. No company may use more than two rigs at any one time.

Companies note the position of existing gas fields on the map.

They can also expect to find gas domes covering more than one square.

Other than that companies back their hunches in deciding which square to place their rig in.

than that companies largely backed their hunches. Gravity maps gave some assistance in indicating the presence of dome areas.

A gravity map of the North Sea can also be used as extra evidence if desired (Fig. 6).

6 High winds have affected drilling, and floating drill ships proved particularly vulnerable to them.

There is a chance factor in each drilling period—the strength of the wind (see 'How to play'). High winds can prevent rigs from drilling, and even (at Force 10) pipelines break.

7 Some companies struck lucky. . . .

An umpire (who has a *Resources* card in his possession) tells players if they have been lucky in their choice or not. The umpire also announces the value of each well that is a successful one.

8 They were then faced with the decision of whether to develop the well or not. Though gas might be present it needed to be there in sufficient quantities to make it a commercial proposition.

There are pipeline costs (based on a figure per square) and companies have to decide whether or not to develop a pipeline. Four pipeline terminals are shown on the map and the pipeline must come to one of these.

The game continues for a certain number of drilling periods—eight is a satisfactory number.

A further step towards reality may be taken by introducing a licensing period at the start of the game. This gives each company the chance to take an option on a certain (say twenty) number of squares before any drilling begins—and to have monopoly rights on those squares in the drilling periods. They choose their concessions in sets of five squares at a time, with free choice of any square. Alternatively the game can be *simplified* by omitting the

Fig 5. Map for North Sea gas game showing concession squares, and the depth of water

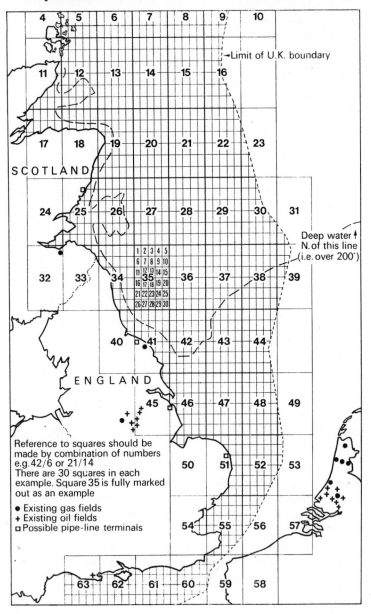

→ Limit of U.K. boundary

SCOTLAND

Deep water ↑
N. of this line
(i.e. over 200')

ENGLAND

Reference to squares should be
made by combination of numbers
e.g. 42/6 or 21/14
There are 30 squares in each
example. Square 35 is fully marked
out as an example

● Existing gas fields
+ Existing oil fields
□ Possible pipe-line terminals

Fig 6. Map for North Sea gas showing concession squares and additional information concerning gravity basins

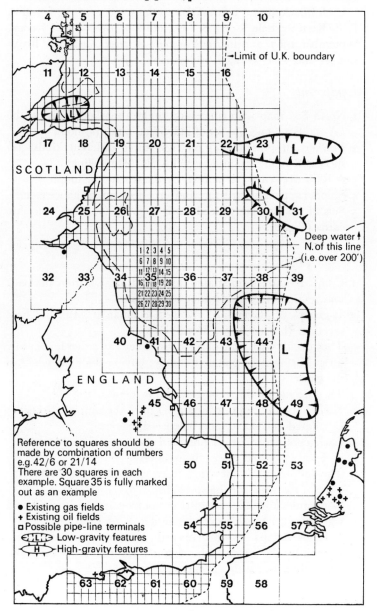

Limit of U.K. boundary

SCOTLAND

Deep water
N. of this line
(i.e. over 200')

1	2	3	4	5
6	7	8	9	10
11	12	13	14	15
16	17	18	19	20
21	22	23	24	25
26	27	28	29	30

ENGLAND

Reference to squares should be
made by combination of numbers
e.g.42/6 or 21/14
There are 30 squares in each
example. Square 35 is fully marked
out as an example

● Existing gas fields
+ Existing oil fields
□ Possible pipe-line terminals
⬭L⬭ Low-gravity features
⟨H⟩ High-gravity features

development of pipelines (paragraph 8). Companies striking gas can automatically claim the value of each square.

Useful background information to the subject can be found in P. E. Kent's article 'North Sea exploration', *Geographical Journal*, September 1967, and in *North Sea oil: the great gamble* by Bryan Cooper and T. F. Gaskell, Heinemann, 1966.

Equipment

1 North Sea map with concession squares marked. This can be (*a*) one large one; (*b*) set of individual ones; (*c*) both.
2 Markers or pins for (*a*) oil rigs—three different kinds, (*b*) successful wells, (*c*) unsuccessful wells, (*d*) pipelines.
3 A Resources card (to be used by the teacher or umpire).
4 Spinner or pack of cards for wind force decision (a chance factor).
5 Sets of (*a*) Company Deed cards (Fig. 7), (*b*) Rig Ownership cards (Fig. 8).

How to play

1 Companies are formed and Deed cards drawn from the pack and retained for the duration of the game.
2 If groups of players are involved, companies form and hold meetings (having drawn Deed cards).
3 They decide what kind of rig to use. They can either hire or buy a rig. Hiring is for one period only; buying is for the whole of the game.
4 They decide by throw of dice who is to go first.
5 Each company hires or buys a rig from the umpire. Costs are calculated and noted on a balance sheet.
6 Each company in turn nominates a concession square (e.g. 35-6, or 47-9) and places their rig on it.
7 The umpire spins the wind force wheel and announces the force of the wind for that period.
8 Companies announce whether or not they are able to drill, and whether their pipelines remain intact.
9 For those companies which *can* drill, a result is announced. This is done by the umpire who consults the Resources card and says either 'Dry well' or 'Gas strike'.

If a company strikes gas they are *also* told the considered value of the strike, per period.

10 All wells are now marked as either 'successful' or 'unsuccessful'.

11 Successful companies can now decide whether to pay the extra costs of a pipeline or not (this is two units per square). If they decide to build a pipeline, the necessary cost is announced and recorded.

No company can claim any benefit from a successful well until the period *following* that in which a pipeline has been built.

12 Then the second drilling period begins. Steps two, three and steps five to ten are repeated.

13 The game continues for an agreed number of drilling periods. Companies aim to maximize their profits.

Strategy

1 It is not always best to hire or buy the biggest and most expensive rig. Its flexibility may be too costly. If a company believes in-shore drilling a reasonable policy a cheaper rig may be used. Similarly a company with low finances may prefer a small-budget policy; grandiose exploration projects are not always advisable.

2 The first strike may be a lucky one. Astute companies may seek to drill nearby (especially if their own resources are low), in order to benefit from a known success. On the other hand, a company with plenty of money may decide that a long-term search in an uncharted area may prove more worthwhile than a scramble for the proceeds of a well-worked area.

3 It may be better *not* to exploit some wells if their value is small. Strikes made far out in the North Sea may not be an economic proposition.

Fig 7. Specimen Company Deed card

VICTOR MARINE EXPLORATION LTD
Scarborough, Yorks.

Starting capital: £2,500,000 = 25 units

Other possible (fictitious) company names: Premier Gas, Meteor Oil and Gas Corporation, Europa Ltd, North Sea Gas Co., Johnson (Texas) Gas and Oil Corporation. Starting capital for companies should be between twenty and thirty-five units.

Fig 8. Specimen Rig Ownership card

Semi-submersible Platform
BLUEWATER

Operates in both deep and shallow waters
Ceases drilling if wind is Force 9 or 10

Costs:
to hire   4 units per round
to buy   20 units (for eight rounds)

Fig 9. Specimen resource card (used by umpire only)

Gas is in the following squares	*Value*
46/30 47/26 47/27 51/2	
27/14 27/15 27/20	
47/23 47/24 47/28	2 units per round
51/4 51/5	
30/6 30/7 30/12 30/13	4 units per round
17/4 17/5	
22/19 22/20 23/16	
41/9 41/10	
63/15 63/20	8 units per round
12/26 18/1 18/2	
47/30 48/26	16 units per round

N.B. This Resources card is based on *real* exploration discoveries in part; if the gravity map is *not* used, an alternative Resources card can be invented for each playing of the game; the company decisions and search procedures are not invalidated.

There should be 4 semi-submersible platforms available; 4 self-elevating platforms; and 2 floating drill ships (for a game with 5 or 6 companies).

Floating drill ships also operate in both deep and shallow water, but cease drilling if the wind is Force 5 or above. They cost 2 units to hire; 12 to buy.

Self-elevating platforms only operate in shallow waters, and cease drilling if the wind is Force 8 or above. They cost 3 units to hire; 15 to buy.

For names, use actual rigs working at the time, or invent a fresh set (reference material on the rigs is fairly easy to obtain from the major oil companies).

7

The Railway Pioneers Game

1 The operational game RAILWAY PIONEERS is one in which players or groups of players assume the roles of railway companies in the USA in the mid-nineteenth century. They build transcontinental railroads from Chicago westwards to chosen goals on the west coast and attempt to maximize profits.

2 Among the aims of the game are:

(*a*) to understand in general the problems that face railway builders e.g.:

to appreciate the importance of taking the physical environment into account (mountains, rivers, passes, etc.);

to highlight the importance of trade and revenue which accumulates from settlements on which the line is built;

to realize that chance factors of various kinds can affect plans;

to understand the effect that competition can have on considered route decisions;

(*b*) to help the pupil's geographical and historical understanding of the USA through familiarity with aspects of its physical and human geography in the course of the game (e.g. to realize more clearly that the land was settled westwards, to discover some of the 'chance' occurrences that affected railways in their building, etc.);

(*c*) to encourage cooperative decision-making and discussion of problems in the classroom as a desirable activity.

Basic ideas

THE REALITY

1 There were several railroads in fierce rivalry in the USA in the 1860s. Each sought to reach the Pacific.

THE GAME

Teams form themselves into Company Boards, and are given a Company Deed with a name. Each member of the team can (if desired) adopt a specific role, e.g.,

Treasurer to keep balance sheet.

Surveyor to plan the route.

Detective to watch the plans of rivals.

Secretary to record decisions.

Chairman to control meetings.

2 The companies sought profitable routes but faced the problems of difficult terrain, and of the need for revenue.

The game is played on a map (Figs. 12, 13) which represents the terrain of the Central and Western USA. The map is divided into squares, and each square is marked with a number. The number represents the assumed *cost* of building through that square. (Thus to build across high mountains is more expensive than across flat plains.) The width of a square equals about 50 miles.

3 They faced the difficulty of bridging large rivers such as the Mississippi and the Colorado.

To build across rivers costs an extra three (units) for bridge construction.

4 They sometimes deviated to include existing settlements

It is possible to gain revenue from squares which are marked with heavy ink and

of the line, from which profitable trade could be gathered (e.g., shipping cattle from the head of the Texas trails).

marked as settlements (e.g. Kansas City, Omaha, Denver, etc.).

Squares vary in cost from two to ten units; squares vary in gain from five to twenty-five units per round.

5 The American continental railroads were built on a land-grant system. For every mile of track built, they were given six miles of country on either side as an encouragement to build. This gave them exclusive ownership of land.

When companies occupy squares, they are considered to own that square. This does not apply to any town squares or to squares within *two* of Chicago. But if a company enters a square *second* it receives only one-fifth of any trade revenue marked for that square.

If one company wishes to build *across* the line of another (possibly because of changed plans or a fear of being cut off) it has to pay ten units to the company whose line is crossed.

6 Various economic crises hit the railroads during building. Some built too quickly with too little capital and were nearly bankrupt as a result.

Companies are given a starting capital on their deed cards (see Fig. 10) plus an income of fifteen units each round (one unit = one million dollars approximately). They can decide how much to spend on building each round, but will also need to keep a reserve of money in their Treasury in case of a chance factor.

7 Chance factors, such as the opposition of Indians, freak climatic conditions,

Two *Chance Factor* cards are drawn out at the end of each round.

disasters, strikes, etc., also pushed up costs.

No company may spend more money than it has available in its balance sheet; i.e., it may *not* go into debt. If chance factors cause debt, companies must cease building until regular income puts them in business again, and miss one month's building operations.

Equipment

1 A map of the USA (west from Chicago) divided into grid squares or sections in relation to the physical environment. This can be (*a*) one large one; (*b*) a set of individual ones; (*c*) both.
2 Set of Chance Factor cards and Company Deed cards.
3 Markers to trace routes; dice and shaker.

It is possible, if desired, to use money for transactions, as used in games such as Monopoly.

If the game is used as part of a teaching theme on railway building and/or the USA transport system, it is also desirable to have a physical map of the USA (with which the grid map can be compared before starting) and a railroad system map of the USA (with which the grid map can be compared *after* the routes have been drawn out upon it). (Figs. 12, 13, 14.)

How to play

1 Companies are formed—either groups of up to five or six or individuals assuming all the roles suggested and acting as a company.
2 Companies draw their Company Deeds and discover their assets.
3 They throw a dice (6 highest) to see who should go first. This order is rotated.
4 Companies hold Board Meetings to decide on strategy. They plan the first period's building programmme (one period is equal to three months' building).

67

5 Each company in turn builds (by counters or pegs) through a set of squares, beginning from Chicago in the first instance.

6 The person who is acting as umpire for the game, or another player, examines the route and checks the cost with the Treasurer of the company. The Treasurer enters this in the balance sheet which he keeps (see Fig. 11).

7 The second company follows the procedures outlined previously in 5 and 6.

8 The other companies do likewise.

9 At the end of each period, the umpire, or one of the Companies, draws two Chance Factor cards from the pack, and announces its news. The factors may affect all companies or only one or two of them. Balance sheets are adjusted accordingly.

10 Income figures are now added in and the second round follows the procedures outlined in 4–9.

11 The game continues until all companies have reached the west coast (having reached the coast, companies can cease building if they wish, and simply draw revenues each round).

12 At the end of game, it is interesting to see which companies have maximized their profits best.

Strategy

1 Companies have to battle not only the physical environment —choosing their route across the Rockies—but also consider the economics of their bank account and the moves of their rivals:

(a) Is it best to press on straight through the Rockies, regardless of cost, in order to get to California first? Or is it better to turn aside and pick up lucrative staging post revenues on the way?

(b) Is it better to spend dangerously in order to build faster than one's rivals, and hope that chance factors are in one's favour? Or is it better to build carefully with a well-stocked bank balance?

(c) Is it better to try to hinder the progress of rivals by 'cutting them off' or deliberately building through squares which they might have used? Or is it better to ignore what others are doing and concentrate on one's own plans?

All these considerations were present in the railroad building which went on in the USA in the mid-nineteenth century, and are reproduced in simplified form in the game. (The chance factors are all based on actual incidents taken from American railroad history books.)

2 The groups of players who form companies will need to talk about the decisions to be taken. A person with rashness or prejudice may handicap a Board; one with vision or persuasion may assist another; a weak Chairman may bring a Board Meeting to ruins, whereas a strong Chairman may be able to reconcile opposing views. And whose voice should be heard loudest on the Board? The game represents some possibilities of discovering how decisions are really taken.

CHANCE FACTORS (to be transcribed to cards). These should be well shuffled and two should be drawn at the end of each round.

1 *Plots.* The American Pacific Railroad and the Great Plains, Rocky Mountain, and West Coast Railroad Company are suspicious of each other's intentions. Intensive spying, illicit bargaining between employees, and perpetual law-suit threats culminate in armed raids by employees of one company on the other. The G.P., R.M. & W.C. succeed in damaging some A.P. track.

Cost: Ten (units) to G.P., R.M. and W.C.

Twenty to American Pacific.

2 *Floods.* A period of heavy rainfall in the Rockies causes damage to bridges. Any company with bridges *west* of a line drawn through Denver must pay for bridge repairs.

Cost: Three (units) for each bridge.

3 *Lack of supplies.* It is difficult to keep regular supplies available to construction gangs working on the line. If there is *no* town on the line within eight squares of its present limit, a company must strengthen its supply organization.

Cost: Ten (units) to each company which needs strengthening.
Delay of one period to companies involved. (Miss a round.)

4 *Indians.* Indians of the Sioux tribe are hostile to builders of northern routes. Any railroad which is building *north* of a line drawn westwards from Minneapolis is hampered by their attacks and must repel Indian raiders.

Cost: Fifteen units for railroad militia, to railroads building *north* of Minneapolis.

Delay of one period to companies involved. (Miss a round.)

5 *Good Weather.* Moderate weather assists construction in the south. Any company building *south* of a line drawn westwards from Kansas City finds that its workers are building faster than they planned.

Gain: Ten units in saved wages, to railroads building *south* of Kansas City.

6 *Labour boom.* A supposed gold strike in the Central Rockies proves to be a fraud. A large number of would-be prospectors have rushed to Santa Fe only to find this out too late. They turn to the railway for work, and are willing to take low wages.

Gain: Ten units in saved wages, to any railroad passing through Santa Fe.

7 *Track quality.* The Government becomes concerned about the standards of building track. Companies must submit to Inspectors if they have built more than twelve squares of track. They pay the costs of improvements.

Cost: One unit for every square of track to each company with more than twelve squares of track.

8 *Bad weather.* Bad weather hits the south. A tornado sweeps across the Great Plains. Any company with track south of St Louis has to carry out emergency repair work.

Cost: Ten units for repairs, to any railroad *south* of St Louis.

9 *Labour problems.* Transcontinental and the Chicago and Western both have labour problems. Workers accuse them of attempting to have 'slave labour', and down tools in strike action.

Cost: Transcontinental and Chicago and Western must cease building for one period.

10 *Avalanches.* Weather fluctuations cause widespread avalanches in mountain regions. Companies which have built through high ground find that the cost of maintaining track is considerable in these conditions.

Cost: Two units for every square of track which has a value of over five (excluding bridge costs).

11 *Labour boom.* Hungry Sioux tribes, forced by depleted buffalo

THE RAILWAY PIONEERS GAME

herds to beg for work, drop their opposition to the railroad.
Any company building *north* of a line drawn westwards from
Minneapolis uses them, and thus adds to its labour force.
Gain: Ten units in saved wages, to any railroad *north* of
Minneapolis.

12 *Economic strength.* The Government becomes worried about
the finances of some companies and orders an investigation of
their bank accounts. Any company holding *less* than twenty
units is put into temporary liquidation and cannot build until
Government assessors investigate further.
Cost: Five units to brief lawyers for Government hearing, to
any railroad with *less* than twenty in balance.
Delay: One period to companies involved.

13 *Disaster.* A relief train runs amok among a construction gang
of the company which is furthest west in building. The gang is
considerably shaken by the experience and depleted in
number.
Cost: Ten units in compensation to workers' families.

14 *Bank worries.* Eastern seaboard bank managers become
concerned about the risk to their money. They fear that
companies are spending too fast and too easily. Any company
holding *less* than twenty (units) runs into difficulties with its
backers.
Delay: No building for one period to any railroad holding *less*
than twenty in balance.

15 *Trade depression.* Companies with towns en route find that a
bad trade depression hits their monthly receipts.
Costs: Town revenues decrease by 80 per cent for the next *two*
periods (i.e. five per period becomes one, twenty-five per period
becomes five).

16 *Cattle boom.* The Chisholm Trail (up from Texas to Santa Fe)
is crowded with cattle. Any railroad which has annexed Santa
Fe benefits from increased trade.
Gain: Five (units) in freight receipts, to any railroad passing
through Santa Fe.

17 *Tornado.* A sudden squally tornado strikes the Mid-West. It
carves a trail of destruction from Omaha due westwards (one

square wide) to the Rockies. Any railroad which has built track in its path has to rebuild it all over again.

Cost: Cost of squares which have to be rebuilt.

18 *Board troubles.* Any company which has built through Denver finds that powerful local opposition is causing difficulties at Company meetings. They have to be 'bought off'.

Cost: Ten (units) for compensation, to any railroad built through Denver.

19 *All quiet.* Nothing happens this period.

20 *All quiet.* Nothing happens this period.

21 *Drought.* Companies building across the high plateau regions find difficulty in maintaining adequate water supplies for their gangs. Time has to be spent on a search for vital supplies.

Delay: Any company building in the plateau regions (between the Sierras and the Rockies—see map) ceases building for one period.

Fig 10. Specimen company deed card

THE CHICAGO and WESTERN RAILROAD CO

Founded 1867   Randolph Street,
                                                          Chicago

Capital assets: 25,000,000 dollars = *25 units*

Other possible names for game companies:

The Great Plains, Rocky Mountains and West Coast Railroad Co.

The Transcontinental Road

The American Pacific Railroad

The New Territories Railroad Co.

Capital assets to vary between fifteen and twenty-five units at start.

Fig 11. Specimen balance sheet

Name of Company_____ Starting capital_____			
	Costs	*Income*	*Balance*
Round 1			
Track building	———		
Effect of chance factors			
Balance	———	———	
			———
Round 2			
Regular income		15	
Revenue from towns		———	
Track building	———		
Effect of chance factors	———	———	
Balance			
			———
			etc.

N.B. It has been found best *not* to allow companies to build branch lines, since this complicates the game unnecessarily.

An alternative end to the game may be played by announcing a closure two rounds after the *first* railway reaches San Francisco. This allows the successful companies to reap the rewards of west coast revenues, and discourages others from deliberately hanging back and not incurring any building expenditure.

A version of this game in kit form now exists. It is published as a teaching unit (Longman Simulation Packs No. 1) and the kit includes a briefing film-strip and a de-briefing wall-chart as well as game materials.

The following three pages show maps which are of value to the game:

Page 74 of the physical features of the area.

Page 75 of a transformation of these into a cost-surface. (This can be used as a board for the game).

Page 76 of the major railroads actually built across the continent, and the order in which they were completed.

Fig 12. Map and profile diagram of physical features of USA

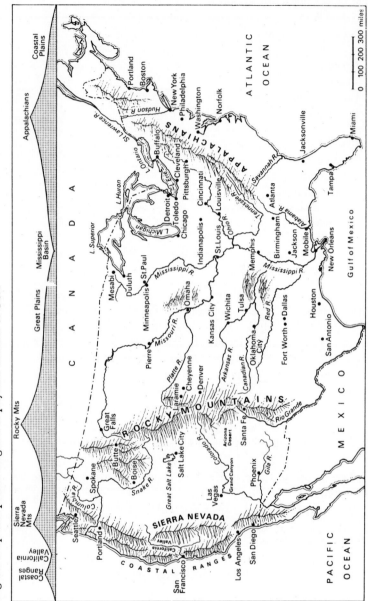

Fig 13. Map for railway pioneers game; a transformation of the map of physical features in terms of cost

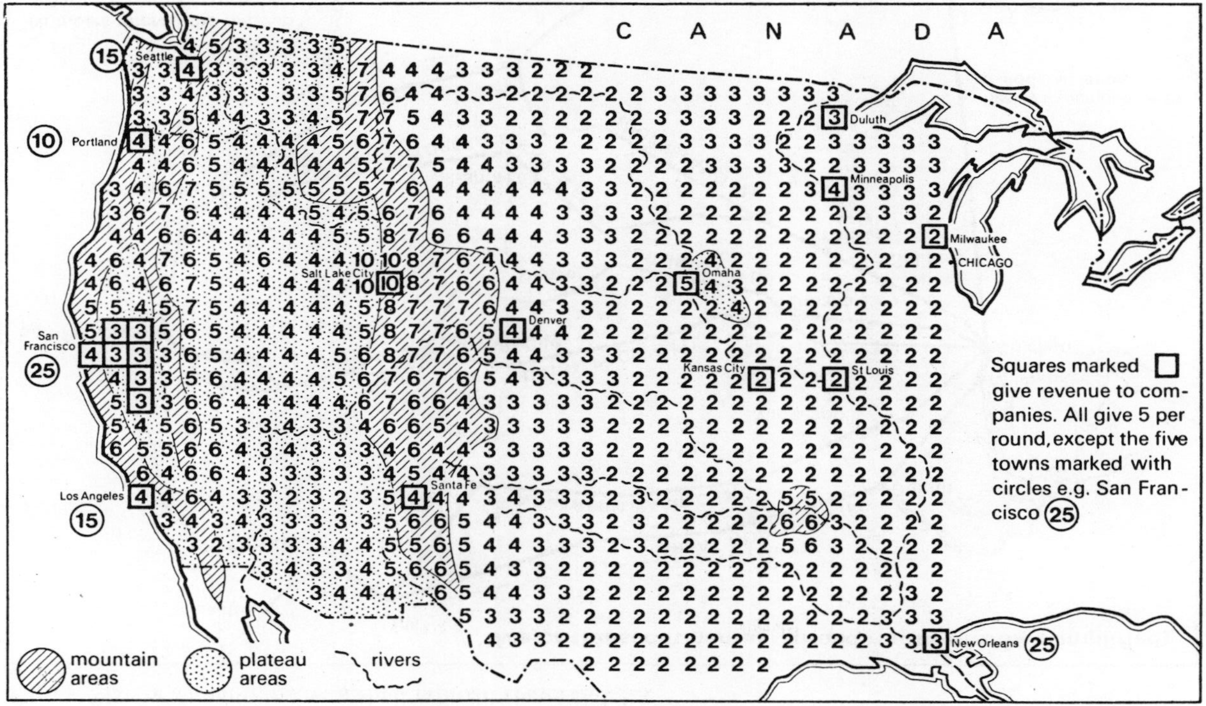

Fig 14. Map of actual transcontinental railroad routes in USA

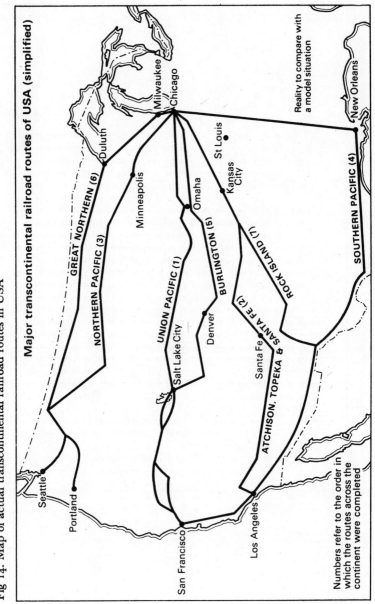

Major transcontinental railroad routes of USA (simplified)

Reality to compare with
a model situation

GREAT NORTHERN (6)

NORTHERN PACIFIC (3)

UNION PACIFIC (1)

BURLINGTON (5)

ROCK ISLAND (7)

ATCHISON, TOPEKA &
SANTA FE (2)

SOUTHERN PACIFIC (4)

Seattle

Portland

San Francisco

Los Angeles

Santa Fe

Salt Lake City

Denver

Omaha

Minneapolis

Duluth

Milwaukee

Chicago

St Louis

Kansas City

New Orleans

Numbers refer to the order in
which the routes across the
continent were completed

8

The Development Game

1 The operational game DEVELOPMENT is one in which players or groups of players assume the role of Development Companies in an undeveloped country. They explore the area by means of chosen methods of transport in search of minerals, and then decide on methods of exploitation when minerals are found.

2 The aims of the game include:

(*a*) to consider the possible methods of exploration and development in undeveloped areas, and to see how these are affected by economic factors and chance.

(*b*) to gain experience in the technique of search procedures and see how this can be done most efficiently.

Basic ideas

THE REALITY

1 In some relatively undeveloped areas, competing companies seek to explore and develop the areas in a profitable way.

This often involves, in the first stage, searching and locating economic mineral deposits.

THE GAME

The game is played on a map (see Fig. 15), which is covered by grid squares. It is based on a West African state, and information about the nature of the vegetation is included (this could have been ascertained by aerial survey).

77

2 Companies usually set out from ports on the coast in order to explore inland, although techniques of air search have had a liberating effect on this pattern in recent years.

Companies can choose to explore in one of three ways in each round:

by rail—in which case they must start from a coastal port and move three squares per period;

by gravel road—in which case they must start from a coastal port and move five squares per period;

by helicopter—in which case they may alight on any square within the state but explore only that square per period.

The cost of these three procedures is assumed to be equal.

3 Though air search is efficient in locating initial economic areas, it is not easy to develop mineral deposits successfully by helicopter alone—road or rail will probably need to be used also.

Companies explore the interior by using one of these three methods. An initial discovery by air may lead to a company building towards it by road and rail, since different modes of transport are able to reap different benefits from each square of discovered resources. If a railway links a square of discovered resources to the coast, the company owning it can claim the *full value* of the square in each round.

If a road is the link, *half the value* can be claimed each round.

If a helicopter is the only link, only *one-quarter* the value can be claimed.

4 The economic value of mineral deposits discovered varies with the size and nature of the deposits.

Squares with mineral resources in them have differing values, which are revealed when a player lands on that square. It is possible to discover OIL, BAUXITE and IRON ORE within the state in commercial quantities.
HYDRO-ELECTRIC POWER sites may also be discovered. There is also a small GOLD MINE—the full value of which can immediately be gained whatever means of transport links it to the coast.

The above represents a simplified version of the game. Paragraphs below represent a development of the game in more complex terms.

5 When developers discover a deposit—they may choose either to exploit the raw material as it stands or (at some cost) develop refining processes in order to increase its value.

Refineries or smelters need to find local power supplies in order to work economically. These local supplies are often gained from hydroelectric power.

Refineries also need links with a coastal port in order to export their processed product.

When resources are discovered players may, if they wish, develop processing plants. This can be done as follows:
OIL. It is possible to build a refinery, but it must be at a coastal port. The refinery is presumed to use its own fuels as a power source.
Cost. A company must forgo its income from any oil square it owns for three rounds.
BAUXITE. It is possible to build an aluminium smelter, but it must be at an H.E.P. site. Companies must discover and own an H.E.P. site before building. The works must be

79

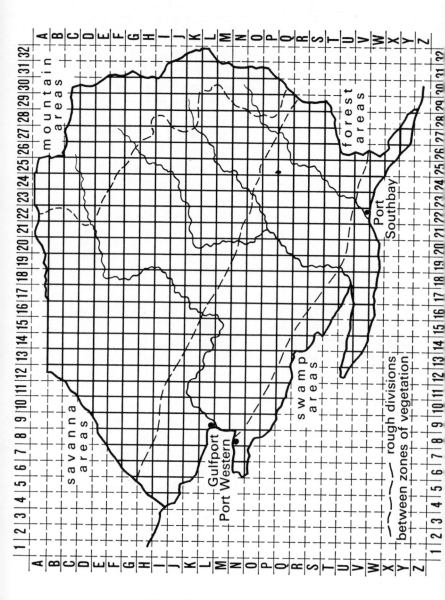

Fig 15. Map for development game

linked to the bauxite deposits and to a coastal port by road or rail.

Cost. A company must forgo its income from any bauxite square it owns for three rounds.

IRON ORE. It is possible to build an iron and steel works but it must be built at an H.E.P. site (as aforementioned). The works must be linked to the iron ore deposits, and to a coastal port by road or rail.

Cost. A company must forgo its income from any iron ore square it owns for three rounds.

6 Processed products are usually somewhat easier to transport than raw materials.

The value of squares *doubles* when their output passes through a refinery. (This double value can be obtained whether or not the link to the coast is by road or rail.)

How to play

1 Players or groups of players form themselves into companies and decide upon their strategy.
2 A dice is thrown to determine who goes first. The first company then nominates a mode of transport (road-rail-air) for survey and moves accordingly, placing markers on the map. Road or rail developments must begin from a coastal port, and can be in any direction as long as they are continuous. Branch lines and roads can be built.

It is not possible to interchange systems. (The presumption is made that internal transfer of goods from road to rail, etc., is uneconomic.) Thus road and rail systems can be developed by the same company but must be considered as separate.

3 The second company does likewise. The other companies follow suit.

4 At the end of the round, the umpire (who should *not* be a player) consults the Resources Card. He tells companies if they have discovered any minerals or H.E.P. sites in their surveys. Once in the squares, companies are presumed to own the rights to minerals of power in that square, though other companies may cross the squares without penalty.

5 The umpire also announces the value of each deposit discovered. This may encourage further development in that area; or the development of supplementary links in a subsequent round in order to realize the full value of the square. Companies note any income received.

6 The second round continues in the same way, with each company being allowed to build either five squares of road, three squares of railway or to survey one square by helicopter.

7 If a company decides to develop refinery or smelting works, it must announce this at the same time as any transport building. It must nominate its site and make sure that this site is connected to the deposit and to the coast. The cost for a refinery or smelting works is to forgo three periods' exploitation income from any deposit of that mineral that the developer owns. *After* these periods it is possible to claim *double* the value from every square of the appropriate mineral connected to the refinery or smelter. A refinery may not be built before a mineral discovery is made.

8 The game continues for a predetermined time. This can perhaps be eight or twelve periods. It is interesting to see which development company proves most profitable.

Strategy of the game

The game poses three problems: by what method to search; where to search, and how to exploit.

The choice in transport into the interior may reflect the outlook of the company. Rail is relatively slow, but a stable basis for exploitation if deposits are found. Road is faster but less able to exploit. Air survey provides small returns on the deposits but may help to locate an initial area more quickly and thus give some purpose to road or rail building.

There are no clues about where the minerals are, and so search procedures must be random. But there are efficient and inefficient ways of covering areas.

If minerals are found quickly, full-scale exploitation by the development of refineries or smelting works may be the right policy. Though this demands expenditure (the forgoing of revenues for three months) it mav pay dividends if the company has large deposits to exploit. On the other hand, a company may prefer the simpler method of taking revenues at their given value and exploiting only the minerals in their raw states.

Differences in policy on these points may well reflect the different attitudes of some development enterprises.

Fig 16. Specimen resource card (used by umpire only)

	Squares	Value per round
GOLD	R 15	£500,000
OIL	G 6, G 7, H 6, H 7, D 22, D 23	£100,000
	H 5, G 8, H 8, L 17	£50,000
BAUXITE	C 25, C 26, D 25, D 26, D 27	£75,000
	B 14, B 15, C 15	£50,000
IRON ORE	B 17, B 18, C 17, C 18, D 17,	£50,000
	R 15, R 16, Q 16, P 16, P 17	
	F 22, F 23	£25,000
	P 9	
	V 18, U 18	
H.E.P. SITES	(Choose appropriate squares on rivers near mountains.)	

Equipment

1 A map of a country (real or imaginary, as desired), marked off in grid squares (see Fig. 15). The game can be played with (a) one large map; (b) a set of individual maps; (c) both.

2 Markers for (a) squares surveyed by road-rail-air;
               (b) mineral deposits; iron ore-bauxite-oil;
               (c) refineries and smelters.

3 Dice and shaker.

4 A resources card (to be used by the teacher or umpire) locating mineral deposits and H.E.P. sites (Fig. 16).

9

The Export Drive Game

1 The operational game EXPORT DRIVE is one in which players or groups of players assume the roles of British exporting companies in the present day and attempt to run profitable companies in the light of the problems of world geography and trade.

2 The aims of the game include:

(*a*) to introduce players to some of the factors that exporters need to take into account, e.g.,

    (i) choice of markets and their capacities;

    (ii) whether it is easy to sell in particular markets;

    (iii) appropriate transportation for particular types of products;

    (iv) chance events which influence the pattern of world trade;

    (v) the policies of competitors; etc.

(*b*) to give some understanding of the 'friction of distance' and its effect in cost terms on different methods of transportation and different types of goods.

(*c*) to teach incidentally some factual world geography (places, countries, oceans, etc.) and to give some knowledge about Britain's principal exports and their markets.

Basic idea

THE REALITY	THE GAME
1 Britain depends heavily on her export trade in order to maintain a satisfactory	The game is played on a large world map. Routes are marked (in stages) from

balance of payments.
British exports are of
several different kinds.
These exports are sent all
over the world.

London to fifteen other
countries in all parts of the
world. Britain exports to each
of these countries.

2 Exporters have a choice of
transportation.
They may use air, ship,
road or rail depending on
the availability of suitable
routes, and their cost and
speed.

Routes on the map pass over
both sea and land.
It is possible to send goods
by *ship* on the sea routes.
It is possible to send goods
by *rail* or *road* on the land
routes.
It is possible to send goods
by *air* over routes which
cross both sea and land.
These methods of transport
have different speeds (see
'How to play').

3 Transshipment of goods is
sometimes undertaken.

Transshipment from one form
of transport to another (at a
cost) is possible en route to a
destination.

4 The kind of transport
chosen often varies with
the nature of the product
carried.

Different kinds of product
have different costs per stage.

5 Among Britain's exports
are:
(*a*) China clay (*b*) whisky
(*c*) precision instruments
(*d*) magazines.
These export products have
different characteristics.

Players in the game run
companies exporting:
(*a*) china clay—a bulky
    product
(*b*) whisky—a fragile product
(*c*) precision instruments—a
    high value product
(*d*) magazines—a product
    which is perishable (i.e.,
    it dates quickly)

6 Companies compete against

There is more than one

each other for overseas markets.

company manufacturing each product, and rivals have to be taken into account in policy decisions.

Company Deed cards specify the cost of transportation for each stage on the map.

7 Markets are at different distances from London and transportation costs differ.

8 Markets also have different capacities.

The capacity of markets is specified on a *Market Information Sheet* issued to each company: warehouses for each product unit are shown on the sheet or map. The sheet also indicates the number of stages to each market.

9 Some markets offer a certain sale; in others, there is much competition and the exporting company may eventually bear the brunt of goods not sold.

It also contains information concerning the certainty of selling goods in each market. Some markets are certain for sales, others highly competitive. (This is determined by the throw of dice.)

10 Unforeseen chance factors affect the pattern of world trade. They may cause companies to alter their objects or policies; they may affect goods en route or future intentions, etc.

As the game progresses, *Chance Factor* cards are drawn out at the end of each round. They simulate possible current events which might occur during a period of trade.

The game makes some simplifications:

1 Each company is assumed to have a constant home market and an equal desire to export.

2 Each firm is assumed to produce one unit of its product in each round. (This unit is defined on the Company Deed card.)

How to play

1 Companies are formed by players or groups of players. They draw Company Deed cards (Fig. 17). The Company Deed cards give them information concerning:

(*a*) the nature of their product;

(*b*) its transportation costs;

(*c*) its costs of trans-shipment;

(*d*) its income per unit.

2 Companies meet and decide on strategy. A dice is thrown to determine who goes first.

3 The first company nominates its destination and its method of transportation (e.g. a whisky company might choose to send its first round unit to New York by sea). It calculates and notes the resulting cost on its balance sheet (Fig. 18).

It places its unit on the board (see Fig. 20) at the appropriate stage of its journey—with the following table in mind (one round = in reality one week, approx.):

Goods travelling by air can move fifteen stages per round.

Goods travelling by rail can move five stages per round.

Goods travelling by sea can move four stages per round.

Goods travelling by road can travel three stages per round.

4 *If* the goods *reach* their destination in one round of travelling time, the question of sale is then considered. The Market Information sheet (Fig. 19) is consulted and the certainty of the market ascertained. A dice is then thrown. If the market is marked four–two, then a throw of one, two, three or four will denote a successful sale, five or six an unsuccessful sale, etc.

If a successful sale is made, the unit is placed in the warehouse for that place on the map or information sheet and the income claimed.

If the sale is unsuccessful, the company has the choice of remaining at the place and trying to sell again next time, or of moving the unit on to another possible market in the next round.

5 The capacity of the market has also to be considered. When the first unit is sold to a particular place, the capacity of

that market decreases by one unit; this will be seen clearly as the warehouses fill up.

6   The second company then also announce their destination and method of transport, and then place their unit at the appropriate place on the board, either in transit or at a destination. If they *reach* a destination then they follow the procedure outlined above.

7   Other companies do likewise.

8   At the end of the round, a Chance Factor card is drawn from the pack. Its effect is noted by each company.

9   At the end of the round companies check on their balance sheets and total costs and income.

It is permissible to go into *debt* during the game (i.e. to have goods in transit with income expected but not yet received), but at the end of the game any income not yet gathered cannot be counted.

10   The second round proceeds as the first; see steps three to nine above.

11   The game continues for a predetermined number of rounds: twelve is a satisfactory number. At the end of that time the auditors are presumed to arrive and the companies' books are laid open for inspection.

The game is not directly competitive between products, but it is interesting to see which one of rival companies exporting the *same* product has maximized their profits best. It is also interesting to compare the kind of revenues made from different kinds of products.

Equipment

1   A world map with routes and stages marked on it. If it is possible there should also be room for market capacities and 'certainties' to be marked, so that product units can be 'placed' in warehouses. (This has *not* been possible on the map attached, but is on a large board.)

There can be (*a*) one large map; (*b*) a set of individual maps; (*c*) both.

2 Sets of (*a*) Company Deed cards (eight or twelve); (*b*) Market Information sheets.

3 Markers for units; dice and shaker.

4 Balance sheets, to be kept by each company.

5 A pack of Chance Factor cards.

Strategy

The game allows a good deal of strategic development.

1 Companies with high-value products, e.g.; precision instruments, whisky, will obviously be able, if they choose, to receive income more rapidly, since they can use faster methods of transport. They must decide whether or not, however, the faster rate of transport is *worth* this extra income. Whisky firms will seek to avoid transshipment during their journey since costs of this operation are very high for them.

In the case of firms exporting magazines, fast transport is essential to far-off places, since otherwise their product becomes dated and unsaleable.

China clay firms, on the other hand, may choose to use slow routes because of their much cheaper transport rates. (Sending china clay by air does not yet seem an economic proposition!)

2 Companies must decide whether to sell to markets close at hand early on (otherwise perhaps their rivals may . . .) or whether to send out early units to more distant markets. If sale to distant markets is delayed too long, faster transport must be used, and chance factors may intervene to prevent the sale being made. Chance factors may also *cause* a company to seek alternative methods of transportation.

3 Companies must also evaluate the risks of highly competitive markets. High capacities may make these attractive, however.

4 A long-term strategy of selling can be evolved, but an eye must be kept on the exporting policy of rival firms, who may steal markets in opportunist fashion.

It is possible for a firm well in debt to have all its income amass in the eleventh or twelfth round and to reap the benefit of cool long-term exporting strategy.

CHANCE FACTORS (to be transcribed on to cards).
Others should be originated as appropriate to the time when the game is played and given a topical context.

1 *London dock strike.* Sea-going traffic is delayed extensively while unions and management negotiate at London Docks.
    Delay: All exports must be diverted *from* sea transport for the next three weeks (rounds).

2 *Russia freezes relations.* Moscow tightens restrictions on British trade as a diplomatic move.
    Cost: Any exports sent to Moscow are taxed—£100 per unit.
    Delay: Any goods en route to Moscow are delayed two weeks (rounds)

3 *Trade gap worsens.* Britain looks as if she is on the rocks again . . . Bank Rate rises; all companies with overdrafts pay extra to the Chancellor.
    Cost: All companies currently in debt pay £10 for every £100 in the red.

4 *Suez Canal blocked again.* Arab-Israeli conflict again erupts at Suez.
    Delay: From now onwards all goods must be rerouted away from Suez, whether sea, land, air or rail.

5 *Imperial Preference Agreements end.* Trade to all Commonwealth countries suffers financially from this decision of the Commonwealth P.M.s' conference.
    Cost: Income *decreases* twenty-five per cent on any export sent to Commonwealth countries in future.

6 *South Africa ostracized.* The Government prohibits all exports to South Africa in a desperate attempt to solve the Rhodesian problem.
    Delay: Any goods bound for S.A. must be diverted elsewhere. No other exports may be made to S.A. in the future.

7 *Devaluation in Chile.* A Government in trouble in Chile is forced to devalue. British exports are able to benefit.
    Gain: Income per unit sold in Santiago increases by 50 per cent henceforth.

8 *New discovery of china clay*. A new rich seam of china clay is discovered in Russia.

Cost: No more china clay can be exported to Moscow.

9 *Pilferage increases*. There is an outbreak of pilferage on road transport in Europe. Any goods travelling on this continent by road are likely to be damaged.

Cost: Goods which travelled by road in Europe in the last week (round) lose half their value.

10 *Prohibition*. A new Indian government decrees prohibition for religious reasons. This affects whisky exports to Bombay.

Cost: Any whisky exported to Bombay in the past week (round) is confiscated and no income is received for it. No further exports allowed to Bombay.

Fig 17. A specimen Company Deed card (inside pages)

Your company produces MAGAZINES They date quickly and so usually have to use fast transport routes (i.e. they are perishable). Any magazine that arrives *more* than three months after its original dispatch will be out of date. It is not sold.	*Starting capital*: £1000 *Production*: 1 unit per week (1 unit = 10 tons of magazines) *Income*: £2000 per unit *Costs*: (Per unit per stage) By air £400; By sea £100; By road £250; By rail £150 Trans-shipment costs: £250 per unit.

All companies begin with the same capital, and produce 1 unit per week. 1 week = 1 round in the game.

A possible set of companies for the game is outlined below:

1. NEWSPAPERS (perishable)
   ABC Newspapers   Information as in Fig. 17.
   British Periodicals Service:

2. WHISKY (valuable but fragile) Unit costs–Air: £200
   Colquhoun Whisky Co.   Sea: £150 Road: £100
   Dalrymple Distilleries: Rail: £150.
                             Income–£10,000 per unit.
                             Transshipment–£500 per unit.

3. CHINA CLAY (bulky raw   Unit costs–Air: £3500
   material)   Sea: £50 Road: £150
   English Clay Corporation.   Rail: £100,
   Fowey China Clay Ltd: Income–£2000 per unit.
                             Transshipment–£150 per unit.

4. PRECISION INSTRUMENTS   Unit costs–Air: £250
   (valuable and specialized) Sea: £100 Road: £200
   Grant Precision Tools.   Rail: £200.
   Halesowen Instrument Co.   Income–£10,000 per unit.
                             Transhipment–£250 per unit.

Fig 18. A specimen balance sheet (partially filled out)

Name of company *Dalrymple Distilleries* Starting capital *£1,000*

Week 1

Decision: we send 1 unit to *New York* by *sea* it will take *3 weeks (10 stages)*

Costs *£600 (4 stages)*

Effect of chance factors ....

Income from sales ....

Balance at end of Week 1 .... *£400*

Week 2

Decision: we send 1 unit to *Bonn* by *road* It will take *1 week (2 stages)*

Costs in Week 2 *£200*
Costs from former rounds *£600 (4 more stages on unit 1)*
Effect of chance factors ....
Income from sales *(Dice throw successful) £10,000*

Balance at end of Week 2 .... *£9,600*

Place	Certainty of market	MARKET CAPACITIES (per 12 rounds)				STAGE DISTANCES				
		Magazines	China clay	Whisky	Precision instruments	Road or rail	Air	Sea (via Suez)	Sea (via Cape)	Sea (via Panama)
ATHENS Greece	4–2	xx	xxx		x	5	5	8	—	—
AUCKLAND New Zealand	6–0	xxxxxx	x	xx	xx	33c	28	32	36	28
BOMBAY India	4–2	xxx	x	xxxxxx	xxxxxxxx x	13	13	19	28	—
BONN West Germany	3–3	xx	xxxxxxxx	xxx	x	2	2	—	—	—
CAPETOWN South Africa	6–0	xxxx	xx	xxx	xxxxxx	20c	17	26	17	—
COPENHAGEN Denmark	5–1	xxx	xxxxxxxx	xx	x	3	2	(2)	—	—
LAGOS Nigeria	6–0	x	x	x	xx	19c	14	—	14	—
MELBOURNE Australia	6–0	xxxxxxx	x	xxxxxx	xxxxxx	29c	25	28	32	32
MOSCOW U.S.S.R.	4–2		xxxx	xx	xxxx	5	5	(7c)	—	—
NEW YORK U.S.A.	3–3	xxxxxxxxx x	xx	xxxxxxxxx xxx	xxxxx	—	10	—	—	10
RIO DE JANEIRO Brazil	4–2	xx	xx	xx	xx	—	12	—	12	—
SANTIAGO Chile	4–2	x	xx	xx	x	—	16	—	16c	20
SINGAPORE Singapore	6–0	xx		x	xx	20	20	23	31	39
TOKYO Japan	3–3	xxxxxx	xxxxxx	xxxxxx	xxxx	20c	20	31	39	32
MONTREAL Canada	5–1	xxxx	x	xxxx	xxxx	—	9	—	—	9

Fig. 19. Each x represents one unit of the product. Thus the capacity of Athens as a market in the total period of the game is two units of magazines, three units of china clay, nil units of whisky, one unit of precision instruments.

As units reach markets they can be marked on this sheet; it will then show at a glance the current state of market capacities.

N.B. Code letter c placed after a stage distance represents a composite route, i.e., one involving transshipment.

The Dover-Calais ferry crossing is considered as part of road and rail routes; not as a sea crossing.

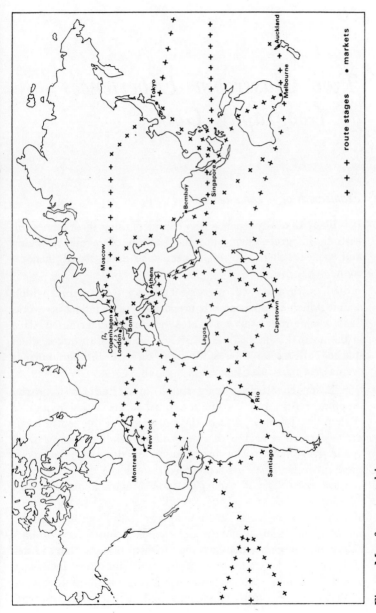

Fig 20. Map for export drive game

+ + + route stages   • markets

10

Two Classroom Experiences of Geography Games

Framework of a game activitiy

It is likely that any teacher who considers using these games will wish to integrate them into the pattern of regular teaching and not leave them as a single spectacular and isolated phenomenon.

Such integration will presume embedding the games within a unit which comprises both preparatory and follow-up work. Such a unit may well come at a previously determined place in the normal syllabus; alternatively it may be introduced when sparked off by an interest brought forward through current events or a particular interest of a pupil.

A lower-school class doing simple local study in a nearby shopping centre may become interested in the actual dynamics of shopping journeys, and in the geography of the centre itself. Perhaps there is a class discussion about whether people shop 'aimlessly' or not. The introduction of the SHOPPING GAME (perhaps using a base-map designed to replicate the local centre in most respects) will then give some insights about the considerations which may be relevant. (The geographical importance of routing procedures becomes apparent through the practicalities of the situation, not through their conscious interpolation.) Following the game, the class may wish to plan or chart routes of shoppers at the centre to see if they took efficient or inefficient routes. They might also go on to consider the distribution of types of shops in the centre to see how this affected the journeys.

In one class where this game was used, the class made most astute suggestions about the use of two empty shops in the central area, based on observed journeys and the frequency with which certain types of purchases were made.

Or, as a second example, EXPORT DRIVE might be used following the study of a table of Britain's annual exports, and a discussion of their possible significance. Following up the game, a class might collect sets of newspaper cuttings to illustrate different kinds of British exports, their destinations and their chosen routes. (The original idea for the game was born while listening to a six o'clock news item in which an exporter explained how he had got his electric fires to Japan via the Trans-Siberian Railway during the 1967 London dock strike.) A total picture of Britain's export market, or that of any other country, might then be assembled with the class sympathetic and better informed about the basic decisions that exporters need to take.

The rest of this chapter describes two actual game operations, with plan, description and comment set out in order to illuminate the practical aspects of operational game-playing in the classroom.

'*Railway Pioneers*' *with fifteen year olds*

Mr Macarthur teaches in a secondary modern school in London. He had a fourth-year class who were possible G.C.E. candidates and who were studying North America as part of their syllabus in that year.

He decided to use the RAILWAY PIONEER game to help him teach some basic factual material, and also to emphasize the importance of the early transcontinental railroads in opening up the West. He decided that he could afford four periods to cover these points.

PERIOD I

He began, in fairly orthodox fashion, one morning by pulling down the large roller map (physical) of the USA which was stored at the front of the classroom. He drew the class's attention to the basic facts which they already knew—the backbones of the Rockies and the Appalachians, which stretched roughly north–south. He went on from there to take up discussion on a TV serial that was currently running, *Iron horse*. This was a Western series which featured Dale Robertson as a railroad owner. The response from the class was enthusiastic about some of the stories seen in recent weeks.

Mr Macarthur allowed the class a few minutes of contribution, and then (having previously consulted sources) mentioned a few of the true exploits that occurred in the early days of the transcontinental railroad building. He spoke of the competition between the routes, of the rich prizes on reaching the West Coast, and of the possible problems. Part of the discussion was as follows:

MR M: What do you think the railroads had most trouble with?

FIRST BOY: Indians, sir—it was always the Indians—they didn't like the railroad. They had a story about that last. . . .

MR M: All right; all right, Hopkins. So why did they have trouble with the Indians?

SECOND BOY: Well, sir, the Indians had never seen railroad engines before. . . .

THIRD BOY: No, that's stupid—they didn't like them, sir, because they thought the railroad companies were going to steal their land.

MR M: But wasn't there lots of land for everybody?

THIRD BOY: Yes, sir, but the Government gave the railways land, didn't they? . . .

And so on.

Mr Macarthur then moved towards the introduction of the game. His important first move was to distribute copies of the game map (see Chapter 7) and to relate it to the physical map of the USA at the front of the room, and also now consulted in personal atlases.

The point which he succeeded in getting the class to articulate was the connection between the costings on the map and the difficulty of the terrain which the railways crossed. He made clear the fact that building across the Great Plains was easier than across the Rockies; and that building bridges across rivers or viaducts across lakes was a business that was costly.

He then went on to draw from the class the ways in which the railways could gather revenue. One boy was an expert on the cattle trails of Texas, and shouted many details about the way in which steers were put on railroad wagons. Mr Macarthur took this information and wove it into his general theme of revenue from towns. He asked the class: 'If you were a railroad builder starting from Chicago, where would you aim for?' Most answered 'San Francisco'. Mr Macarthur gently pointed out the difficulties of the terrain if a straight route was taken. 'Well, sir,' said one boy, 'the shortest route isn't necessarily the quickest, is it—or even the cheapest?' 'A good point,' said Mr Macarthur. 'Now let's see if you can work that out in practice.'

Then, amidst some whistles of surprise, he announced that the class was to play a game. He produced a pegboard map of America and related it to the atlas physical map. The map bore some resemblance but was patterned with holes for plastic pegs. He spent five minutes organizing his class into five companies, and distributed Company Deed cards to them.

He explained briefly the procedure of playing the game, pointing out the importance of remembering the Chance Factor cards, of conserving finances, and of noting what the opposition were doing. He then told the five groups to appoint their officers in the few minutes left of the lesson. He suggested:

(a) a Chairman (to keep order);

(b) a Secretary (to take notes of the meetings);

(c) a Treasurer (to keep the balance sheet, distributed with the deed card);

(d) an Assistant Treasurer (to check the Treasurer's maths);

(e) a Company Surveyor (to study the map in detail and make suggestions about which route to take);

(f) a Company Detective (to keep an eye on the other groups' activities);

(g) a Company Engineer (entrusted with job of coming out to the main board and putting the pegs in the right spot).

'We shall draw for who goes first,' said Mr Macarthur. 'Winners can move today, second team tomorrow, and so on. Come in at break time while I'm here and I'll help you with the first move. If you forget to come then you lose your move.

Then by next Tuesday each company will have made an initial move'. (This was a wise plan since it enabled each group to deliberate fully before their first move; if this is done in class time there are sometimes hold-ups as preconceived strategies are wrecked by the moves of those who go earlier.)

The class went out chattering excitedly—no group forgot to come and make their moves in the intervening days. Three groups came almost *en bloc* and spent time discussing the game further with Mr Macarthur; one group sent two boys (who argued fiercely before placing their pegs); one group had a Company Engineer who quietly came and got on with it all alone, briefed by a grubby note from his Surveyor (the brains of the company).

PERIOD 2

'Now,' said Mr Macarthur, as the lesson began—and the class settled more quickly than usual—'let's have a look at these first moves, shall we?'

He took the pegboard with the five different lines of pegs in it, and made some brief comments. He noted that Group A had set off straight across the board, but that both Groups B and C seemed to be sliding southwards. Group B having elected to spend more money than C and so having a slight lead, had already captured St Louis. Group D was halfway between the middle route and the southern route but had been cautious in its expense. Group E, the fifth to start, had taken the other obvious option, and swung north to Milwaukee and then on in the direction of Minneapolis.

The chairs were pushed into groups as the Company Boards discussed their next moves (already some had made visits to see how the board was progressing). Mr Macarthur allowed five minutes for discussion before Round 2 was begun. Some quite heated discussion began to develop in Group D (whose strong-willed Treasurer counselled caution) and in Group C where a change of route policy was already being canvassed.

Then, as Mr Macarthur called them out one by one from lots drawn again, the Company Surveyors came and placed in further sets of pegs. Mr Macarthur checked the finances of each team by asking them to show him their balance sheet. He noted that Group B were fairly reckless. When he asked two boys to draw Chance Factor cards from the pack there was a breathless hush as they read out the factors and the penalties.

Then there was renewed discussion for a few more minutes; the room was fairly noisy, but Mr Macarthur ascertained by a quick stroll round that all the noise was quite clearly devoted to ardent discussion of the problem in hand. He noted that the slower boys had usually opted for the job of Company Engineer or Company Detective; he pointed out to one group that this did not mean literally overhearing other people's discussions but noticing what the other strategies were on the board and pointing them out in his own group. He also noticed that one or two personalities began to become prominent in each group; he stopped the class at one stage to point out that everybody had a right to be heard and that the Chairmen of the companies should take votes about which routes to take and not be tyrants. Said the Chairman of Group B: 'But it would be better if I *was* a tyrant, sir.' Amid the howl of protest from the rest of his group, Mr Macarthur observed that he had a little to learn about making friends and influencing people.

By the end of the lesson three more rounds had been played. By this time Group E was firmly committed to the northern route, Group A was progressing well through Omaha and the central areas, while Groups B, C, and D were by now locked in mortal combat—all apparently attracted by the magnet of the Santa Fé Pass.

Group B, now dominated by a lively and personable Company Surveyor, had reached Santa Fé itself, ahead of Groups C and D, who had suffered delays and disadvantages through the turn of the Chance Factor cards; but B were running their finances dangerously low in order to achieve this lead.

At the end of the lesson, Mr Macarthur announced the weekly homework. 'I want each group to submit a report to me of the decisions they have taken so far, and the reasons for them,' he said. 'It can be a joint effort but it ought to be written by one person and agreed to by the rest.' The five reports were duly received two days later; one had a 'minority report' attached from a dissident sub-group in Group C who were unhappy about the money being spent on strong competition, and who wanted to see their company go even farther south.

PERIOD 3

At the start of this period Mr Macarthur had to deal with a quite serious problem. Unknown to him, there had been a private argument between two of the groups. The Treasurer of Group B accused a member of another Group of taking his balance sheet out of his desk and looking at it; the other pupil (in Group A) vehemently denied this, and said in defence that he had found the sheet on the floor and couldn't *help* looking at it. Mr Macarthur smoothed things over, and privately noted the lengths to which company competition was going. He took the steam out of the problem by reminding both groups that friction of this kind could cause the abandonment of the game if it was not checked; they took the point.

The same procedures were followed as in Period 2, excitement mounting as Groups B and D were seen to be nearing San Francisco together. Pride went before a fall, however, for Group B. In their first round of this period, they again spent lavishly, and then were penalized severely by a Chance Factor card which called for rebuilding of early stages of track owing to bad weather. They went 'into the red' and were forced to halt for a month. In group C, now falling behind, another crisis developed. The Chairman came out to Mr Macarthur and said he was resigning since he was in a minority of one over the speed of building. Mr Macarthur accepted the resignation, and asked him to remain with Group C; the others in the group agreed to this, but a new Chairman took over, and immediately encouraged the authorization of faster building. The predecessor shook his head sadly, but was outvoted.

Up in the north, Group E got on quietly with their building and arrived in Seattle, the first group to reach a West Coast terminal. Other groups suddenly became alarmed as Group E then turned south and began to build quickly towards San Francisco from their far northern domains.

Eventually, amidst scenes of high excitement (held firmly

in check by Mr Macarthur) Group A just reached San Francisco first, but could not prevent Group E from extending their lines into many areas of the Golden Valley. Group D reached Los Angeles, while Group C finished nowhere in particular, reaping the reward for excessive caution. Group B failed in their objectives also, but their Surveyor explained things away to his fellow-members with much good-humour and inventiveness: 'Look, we've had bad luck—but I'm sure that, if we have the chance to build railways across Africa or South America in the near future, our policy will pay off.'

There was just time, before the bell went, for Mr Macarthur to point out that the balance sheets would reveal the important state of affairs, and to ask Treasurers to check these carefully before the next lesson.

PERIOD 4

Mr Macarthur began by asking the Treasurers to let him have their balance sheets. From these comparisons, he announced that Group E (with the far northern route) had the most favourable balance, and was the most economically sound railroad company. In subsequent discussion, he pointed out that Group A had spent heavily in tackling the Central Rockies head-on, even though they reached San Francisco first. Groups B, C, and D had spent a good deal of their time gently blocking each other (one even to the extent of building a crescent-shaped route out of Santa Fé in order to make another company have to cross its lines and thus pay out more money); their scrapping had prevented any one of them from reaching the West Coast easily.

He then pulled the game back to reality, and talked about the history of railroad competition and how this game had shown up some of the strategies which had occurred in the U.S.A. and in other parts of the world too. (He mentioned the deliberate 'blocking routes' built in England in the 1850s and 60s, instancing one local example.) Then he produced a map of the railroad systems of the United States, and asked the class to compare the actualities with their own game developments.

With some surprise they discovered that the routes of the

Great Northern Railway had been followed almost exactly by Group E, the Union and Western Pacific line had been roughly followed by Group A and the Atchison Topeka and the Santa Fe line had been approximated by Group D (except for a failure actually to get into Santa Fe). Mr Macarthur spoke of the real-life success of the apparently unpromising northern route and of how Jim Hill had realized the possibilities of this extra route, and capitalized on its physical and economic advantages.

He also spoke briefly of the other routes across the continent, and pointed out the physical factors that had influenced them. Then, taking up the pack of Chance Factor cards, he reminded the class of what had already been drawn out. 'What else do you think is in this pack?' he asked them, 'What other factors might we have come across?' As the suggestions came—the problem of supplies, of droughts, of disasters, he riffled through the pack and read out the appropriate cards—almost like a magician finding the right answers. The class couldn't wait to put their hands up and offer more suggestions—their motivation was of the highest order.

As the end of the lesson approached, Mr Macarthur announced the second homework connected with the topic. 'I want you to write the story of *your own* part in the railroad building,' he said. 'You can be factual and simple; or else you can dress it up as a really colourful story, if you want—that's one part of the homework; the other part is to take some more of these maps (here he distributed another game map to each boy) and from your atlas trace the routes of the real companies on to this map. You can put some explanation underneath if you want to.'

AND AFTER . . .

It appeared that they 'wanted to'. When Mr Macarthur looked at the books in the following week, the explanations of the real system amounted to two or three pages of writing in some cases. The biographies written ranged from adequate factual one-page accounts of what happened during the game to an eighteen-page 'Western style' story, written by one boy with graphic detail about the game, the characters on his board, and colourful incidentals which he had dreamt up in fairly faithful style.

Two weeks later, Mr Macarthur found one boy reading the history of the Union Pacific railroad; another four began to devise their own game, based on explorers crossing the Sahara; another group sought and gained permission to come in and play the game all over again in the period before five p.m. when the caretaker locked up. They played almost every afternoon for two weeks.

At the end of term, Mr Macarthur asked the class if they had liked the game. He asked them to write down their reactions on a sheet of paper. Some of the random comments are given below.

I thought that it was going to be a bit stupid at first, but I found it amazing to see how we went the same way as the builders of the past. It really made you think.

I enjoyed it very much—geography lessons have never been made so interesting before. We learnt much more from the game than from anything else this term.

It was very useful for revising the physical geography of America. I now *know* where Omaha and Denver really are.

I am not sure that we needed to spend all that time on the game—but it was a very good change from writing notes. . . .

You have changed geography for me, sir—I now think of it as an exciting real-life subject. More games please!

The game taught us a great deal about the way in which railways are run. It was most interesting to try and act like a railway company—I enjoyed the arguments on our company, and especially enjoyed my job as Treasurer. My father is an accountant and he helped me at home; he was surprised at what we were doing but seemed very pleased.

I think that the Chance Factors were perhaps too powerful; they handicapped unfairly. But then that may be life as it is, I suppose.

These games are a good idea. They are quite different from ordinary teachers [sic]; I don't think we should have them all the time, but they certainly wake you up. Surprisingly, though it was enjoyable we were learning at the same time.

There is one interesting tailpiece to all this. The winning

group was chaired by a rather quiet boy. Perkins, the moving spirit of Group B, came up to Mr Macarthur after school one day and said; 'I can't understand it, sir—you know he isn't meant to be any good at geography—at least not at the normal geography that we do.' 'Well,' said Mr Macarthur looking him straight in the eye, 'I suppose it all depends on what you mean by geography, doesn't it?' Perkins, who always came top in the 'normal geography' tests, gravely nodded his head and went away to ponder the matter.

'North Sea Gas' with eleven-year-olds

The following set of notes, based on playing NORTH SEA GAS, are taken from a teacher's workbook. The school was a comprehensive school in inner London, and the class (1D) a large one; thirty-six, eleven- and twelve-year-olds, ten per cent of them immigrant children of the first generation.

WEDNESDAY, PERIOD 6

Aim. To use a geographical model so that the class begin to appreciate the underlying principles of some of the geography of Britain's energy.

Methods.
(a) I shall talk to the class about different kinds of energy—coal, gas, H.E.P. etc. (ten minutes);
(b) then I shall show them the material that we have gathered from B.P. and talk about the recent developments in the North Sea (ten minutes);
(c) I shall go on to look at the concession map, and relate it to the base board which I have prepared.—We will discuss where oil or gas might be found, but I shall not introduce the gravity map as I think this complication is too difficult at this stage (ten minutes);
(d) I shall divide the class into groups for the game, and make sure that they understand major points, e.g., different capabilities of the three kinds of rigs; how to keep balance sheet, etc.

Comment. The class were very excited when I pointed out that a game could help us with geography. They knew a good deal about oil rigs already (several watched 'The Troubleshooters', a TV programme which dealt with this subject).

We had groups organized by the end of the period, with chairmen and treasurers appointed. Many children stayed behind to look at the maps at the end of the period; June Spriggs pointed out that rigs could not drill in bad weather—I told her we would consider this as we played the game by bringing in the factor of wind force.

FRIDAY, PERIOD 8

Aim. As above—see Wednesday's lesson.

Method. We shall spend this lesson prospecting for gas. I am not sure how long each round will take, but reckon that we may get through three or four rounds—somebody should strike gas by then.

Comment. I overestimated the children's ability to complete the transactions; we completed only two rounds. The lesson was very lively and the children were really excited. I am afraid we disturbed the sixth form next door by a loud cheer when one group struck a well. I was so engrossed in the lesson that I did not let them out till five minutes after the end of lessons and that on a Friday afternoon! They seemed quite sorry to go—unbelievable! Brian Rapley's group struck gas first, just off the Norfolk coast. He told me afterwards that he had drilled there because 'his Dad had told him that was where some gushers had been found'. I fear Brian Rapley's Dad is confusing his oil with his gas, but obviously reads his newspapers (Q. Should the game have the gas in the actual places, or should they be imaginary—if the former, diligent searching can provide the right answers out of game-time. Will query this with R.W.) Eric Oliver and his group immediately drilled next door to Brian's group and also struck gas; Brenda Maghee and her friends have banked on semi-submersibles well out in the North Sea—the others are mostly prospecting inshore with rather more caution.

I have abandoned the pipeline idea in the game; it may be

O.K. for older children, but it will make it too difficult for these first-formers. They are having trouble with their accounting as it is. I will try to prepare some proper balance sheets for them over the weekend—keeping the figures on scraps of paper admits too much error. Instead of building pipelines I shall allow them to count the value of the well and add it directly to their account.

WEDNESDAY, PERIOD 6

Aim. As above.

Method. We will have two, perhaps three, more rounds of the game today. I shall allow the topic to carry over until Friday and then set a homework on it.

Comment. Now that rules are better understood, we have had a more organized lesson. Instead of getting the group to come out and put their pins in the map I allowed only one person to come; this made things much easier. We did three rounds in this period, and there were five minutes left at the end for me to talk to them about how they were getting on. I have said that we shall have two more rounds on Friday before we finish. Several have been coming into the room to look at the map and discuss what to do next.

FRIDAY, PERIOD 8

Aim. To complete above.

Method. Two more rounds of the game, and then a tally of the finances, and a round-up discussion.

Comment. Very exciting finish to the game. Brenda Maghee's group finally struck a rich gas vein out in the North Sea. When I consulted my card and said 'strike', Brenda gasped and went fluttery, as if she'd won the pools. This late strike brought them reward for not clustering around the smaller strikes offshore (where most of the others were prospecting).

John Silver exclaimed to the rest of his group, 'I *told* you those floating drill ships were no good—but you wouldn't take my advice,' just like a businessman.

Checking the balance sheets took a little time—I found a

couple of errors in the additions and subtractions, despite having given out some typed 'balance sheets' that I prepared on Thursday night. (Q. Are there too many noughts in some of these figures—will query with R.W.)

At the end of the period, I calmed the excitement a little—it was necessary after that finish—and asked them why they thought we had played the game. After a little while, the facts seemed to dawn on them. One or two were worried that it was not exactly like the real thing, but I pointed out to them that it was the actual operation that was interesting, with the choice of rigs, the weather factors, etc. I mentioned the gravity map to them, and said that in fact the actual companies had a bit more to go on than they had had, but that it was still a very chancy business.

I finished by talking about the probable conversion to natural gas in many homes, and how this would mean that the girls would have to learn new techniques of cooking. We began to get technical about the difference between ordinary gas and natural gas but I told them to ask Mr Blake (the science master) to explain further.

Homework set. Using the B.P. pamphlet material, and any information from newspapers at home, etc., also any information gained from the game, write about the recent developments in gas exploration in the North Sea. Our textbook does not mention it as it was printed in 1962.

11

Building Games and Simulations

The prospect of building and developing one's own games or simulations may seem a daunting prospect at first sight. But the task can be reduced to realistic proportions if a sequence of development is identified. The sequence outlined below is not intended to be definitive, but it represents at least one reasonably satisfactory way of setting about the task.

In order to outline the sequence more clearly the steps and associated comment are also related to the development of a particular game. This game, 'Urbanization', is not outlined in the text, but it may be possible for those interested in playing it to put it together themselves by the use of this do-it-yourself simulation kit.

Step 1 Identify the basic concept or process which it is desired to illuminate

Example. During an urban geography syllabus, it may be considered important for students to understand *how* rural areas become urbanised, and to look at the forces and strategies involved, i.e., to throw light on *urbanization*.

Step 2 Define the context and scope of the simulation

Example. If urbanisation is decided on in Step 1, the .*context* for the simulation may be defined within a typical English rural area which has scattered settlement, and to which the forces of urban development are being attracted. The *scope* of the simulation may be defined within the area of a single hypothetical (or actual) parish in which there is only one major settlement, which thus becomes the focus for the simulation.

Step 3 Locate the nature of the proposed simulation in respect of
(a)   *Practical equipment;*
(b)   *Personal contribution;*
(c)   *Competitiveness.*

These three characteristics may be best represented by axes, as below;

|————————————————————————————|

(a)   No equipment   Full equipment
       (e.g., Role-play, discussion)   (Each process simulated
                                                              with materials)

|————————————————————————————|

(b)   Full personal contribution   No personal contribution
       (Players free from any   (Players bound by predetermined
       predetermined rules)   rules in all respects)

|————————————————————————————|

(c)   Full competition   Full co-operation
       (Each player operates against   (Players operate as one
       others)   unit)

Example. An urbanisation simulation might be designed to operate on a board, marked out in one-acre plots. This introduces some equipment into the plan. It might be designed however so that real or 'game' money was *not* used in financial transactions, these being carried out only through calculation on balance sheets. This would exemplify the restriction of equipment use. Thus the simulation might come about halfway on axis (a) above.

The simulation might operate with a set of determined rules for players; but within these rules there might be room for players to adopt different strategies and approaches. Thus again the simulation might be placed halfway along axis (b).

In some existing commercial board games (such as Monopoly)

each player is in competition with all the others. This would place such games at the extreme left of axis (*c*). In an urbanisation simulation, however, it might be thought that elements of both competitiveness *and* co-operation might be apparent in the development of the chosen environment, and these elements would thus be emphasised. This would move the simulation farther right on axis (*c*).

Step 4 Identify the participants in the simulation

Simulations need the contribution of differing interests and parties. Within a class of students the following situations may occur:

(*a*) A group of students may simulate a particular person;

(*b*) A group of students may simulate a group (for example, a company, government board, etc.);

(*c*) A single student may simulate a particular person;

(*d*) A single student may simulate a group.

Example. In an urbanization simulation, it may be decided that the participants are a mixture of groups and individuals;

Groups: rural district council, industrial concerns;

Individuals: Local squire, local landowners, private developers of houses.

These participants need to be identified so that they may later form the basis of groupings in the classroom.

Step 5 Define the objectives of the participants

Such objectives may be simple to express (as financial reward) or more complex (such as 'happiness', 'success without unpopularity').

Example. In the urbanization simulation all could be said to have 'satisfaction' as their objective, but to have different ways of defining this.

Thus a further analysis of objectives might be necessary. The local squire would have the preservation of the environment high in his priorities and be relatively unconcerned about profits made on a piece of land; an industrialist might well be said to see his satisfaction in a healthy profit balance on his factory, rather than in giving it a desirable environment.

Both these views might not be completely opposed, however, and thus some room for manoeuvre between such participants is possible. Thus . . .

Step 6 Define the interaction process in the simulation

(i.e. What is the process that is going to *happen?* In what way will the model *work?*)

Example. In relation to an urbanization simulation, it might be assumed that a local squire will sell some of his land to an industrialist or a developer *if* he needs the money badly enough and *if* he feels it is not damaging the environment beyond a point which he cannot bear. The urbanization process thus takes place through a whole set of transactions in which existing rural owners sell land to developers for a changed use, most usually factories, houses or shops.

This process will therefore form the centre-piece of the simulation in action.

Step 7 Transform participants, objectives and interaction into a game reality

This step in simulation building is the most crucial, and the one on which most time is usually spent. Steps 1–6 now need to have a practical and simplified structure given to them.

Example. In the urbanization simulation suggested, the interaction process could be closely and accurately simulated by a system of bidding for land carried out within the confines of the classroom by various groups of students representing participants. The simulation process could further be developed by dividing up the process of bidding into:

(*a*) Oral negotiation—students talk to each other informally;

(*b*) Written bidding —students commit offers to paper and accept or reject these bids.

Thus, both the flexible 'bargaining' and the legalistic 'bidding' side of property deals are introduced. A *round* in the game might

GAMES IN GEOGRAPHY

consist of the oral and written sessions taken together, and be considered to be the equivalent of *one year*.

It might further be suggested that the interaction was quite free between all players, but that the group representing the council should have a defined function of recording (or even agreeing to) all decisions.

Translating the objectives into game reality may seem more difficult at first glance but is not impossible. Financial satisfaction can be defined in terms of money but how to define 'satisfaction with the state of the environment'?

One way might be to suggest that a 'state of the environment' index could be calculated on the board at any time by:

(a)   noting the total number of squares developed;

(b)   noting how many *outside edges* of squares could be counted. Thus if an environment (on the board) was both heavily developed, and 'patchily' developed (with large numbers of individual squares *not* adjoining each other), the 's.o.e.' index would be low. Limited and more coherent development would not push the index down so much.

Even so, with a way devised of calculating the 'state of the environment' in game terms, the problem of evaluating this for *different people* remains. This might be resolved by ruling that participants multiplied the 's.o.e.' index by different strengths, in adding it to their score of 'satisfaction'. Thus, the local squire—who rated it highly—would multiply by ten, but the industrialist only by two.

In ways such as this a potentially difficult problem can be reduced to a reasonable game reality whilst preserving much of its original character.

Step 8 *Add constraints and framework to the game reality*

Once the central part of a game has been defined, as in Step 7, it remains to 'tidy it up' and make sure that it is in balance. Constraints may be needed:

(a)   to simulate constraints in reality;

(b)   to keep the game playable.

Example. In an urbanization simulation, it would be necessary to ensure that the unit figures set out for 'satisfaction' and for 'financial reward' balanced in a reasonable way and were not wildly out of proportion to each other.

Constraints needed to simulate constraints to the situation in reality might include:

(a) the taking into account the need for planning permission—this might give the council group a key role in the game;

(b) the possibility that council policy might change with a different political party in control—perhaps simulated by spinning an *election coin* every three rounds (years) and having a twin set of roles allocated to those who are playing the council group;

(c) the need for essential services to be provided to new pieces of development—perhaps developed by making locational restrictions on development in relation to roads, and by making the council the provider of such services.

N.B. These constraints add greater reality to the model but they inevitably make it more complex. If such extra rules are needed in large numbers in order to make any sense out of the basic situation, the simulation can obviously have a more limited use than a simple one. This in itself may be a good indication for the suitability of a simulation with any age range.

The development of the role of the council is a fairly important factor in the present British urbanisation process. The simulation projected above will perhaps be more successfully used with, say, fifth or sixth formers than with first formers, because of the complexity of the *idea itself*. It would be possible to use the simulation with much younger students, but only in a much emasculated (and therefore less realistic) form.

Constraints needed to keep the game playable might include:

(a) limiting the number of bids and counter-bids to a certain number in each round, in order not to allow interminable bargaining to slow up the general flow of action. Such limitation would itself have an analogy in real life;

(b) suggesting a designated size for factory development, so that the industrialists were seeking to purchase larger parcels of land than the house developers—such a constraint would help to clarify the slightly different policies of the two groups.

Step 9 Draw up the rules of the game

Armed with the work done in Steps 7 and 8, the developer can now set out a 'How to play' sequence, and check his arrangements.

Example. In the urbanization simulation described above, this would involve ascribing roles to players, agreeing on the length of the game, and then indicating the sequence of:

(*a*) oral negotiation;

(*b*) written bidding;

(*c*) recording of transactions made. An agreed finishing sequence would also be set out.

Step 10 Compare the developed game to reality

After all the work done in previous steps, there needs to be a look back to see if the process of simplification and game transformation has still preserved the basis of what was intended for study. If not, the simulation must be at least revised, if not rejected.

Example. In the urbanization simulation, the developer of the game looks back at his objectives, interactions, and rules to see if they will illuminate the kind of urban process he considered in the first place. (In some delightful and rare cases they do so even to himself.)

The simulation outlined above as an example to the ten steps of development is a relatively complex one for school work. It was used, however, in order that some of the difficulties of complex simulations could be shown.

In many cases, simulations need not be as complex as 'Urbanization', and most of the games previously outlined are not as difficult to design.

It would be a mistake to conclude, however, that simple simulations and games do not need the same *rigour* in development. It is vital that the educational simulation should remain close to the reality it seeks to illumine, and not simply enjoy the vitality of its own structure and interest.

Appendix A

Notes on games materials

BOARDS

As indicated in Chapter 3, experience seems to show that boards which can be exhibited in a vertical position are more useful than those which can function on a horizontal surface. Among the possible types of these are:

1 Thick cartridge paper (or large-square graph paper) pinned to existing pinboarding.

   If this method is used, map pins of various shapes, sizes and colours can be used as symbols. These are available from many local stationers, but a comprehensive supply can be found at Edward Stanford Ltd., 12–14 Long Acre, London, W.1.

   An alternative method is to cover the cartridge paper with a large sheet of acetate, that can be written upon with Chinagraph pencils or felt-tip pens, and then washed clean or disposed of after the game.

2 Pegboard (or any type of hardboard with a regular patterning of holes).

   Plastic pegs, of many different colours, are available from E.S.A. or Invicta Products Ltd., Oadby, Leicester, cheaply. They are often used in schools by timetable experts!

3 Ordinary hardboard. This provides a good firm surface, though there are some difficulties in using water-based paints on its surface.

   Adhesive materials can be cut to shape, for use, or possibly bought already cut. Of these Sasco (obtainable from Edward Stanford's) has proved to be most durable; it sticks firmly but can be peeled off with ease.

Plastigraph is a similar material, also available from Stanford's.

4 Polystyrene sheeting (half-inch thick) can also be used for a game surface. It is easy to use with map pins, but has two disadvantages:
   (a) increasing use of the surface leaves its indelible mark in small holes;
   (b) it is fairly fragile, and therefore is best used with a hardboard backing.

5 A magnetic board surface.
   Counters can be made to match. The method is very effective but relatively expensive.

6 My own earliest experiments were with tracing paper, but this is not really suitable for total class viewing, although usable when small groups are playing a game.

REPRODUCTION OF FIGURES

The illustrations of game boards in this book can be used in a straightforward way as bases for classroom games.

If this book is placed under an *episcope*, and the lens is directed towards a wall, paper or card of the requisite size can be placed in line with the directed (and enlarged) image. The game base can then be traced off directly, with the minimum of trouble and inconvenience.

OTHER MATERIALS

Most of the other materials can be easily made from scraps that are found in any stock-cupboard.

Deed cards, Chance Factor cards, Shopping List cards, etc., need only small pieces of card, postcard size, for their base material.

Money can either be used in a theoretical way (by transfer of figures on balance sheets) or, if required, by the use of game money. I normally use money from old sets of Monopoly, counting the numbers on the notes as units of £1,000 or £100 as appropriate to the game. Game money can be purchased from Barnums (Carnival Novelties) Ltd., 67 Hammersmith Road, London W.14.

If a teacher is using the more complex games, access to a duplicator is necessary, in order to provide the accompanying material for each group. It is not necessary on the whole for each group to have a set of rules in their possession, although everyone should have access to the master copy for reference.

Appendix B

Some suggestions for other games

There are many other situations, besides the six set out in the text of this book, in which game situations are inherent.

For example, the morning journey of the commuter in which he often leaves home at the last minute and strives to be in time at his office has much of the 'game' about it—although perhaps it may not seem so when actually being performed! The commuter may be faced with the possibility of alternative routes, of unexpected delays through a variety of reasons, and with significant changes in traffic flow. He seeks to minimize his journey time by successfully evaluating and overcoming these factors. COMMUTING can well be developed as a game approach to the traffic problems of towns.

RAILWAY PIONEERS represents, in this book, an approach to routing and transportation problems. Games could be devised with equal ease to simulate competing airlines or steamship companies as examples of this. Each would involve a similar investigation and analysis of transportation situations.

The farmer is also in a perpetual 'game situation'. He seeks to maximize his profits and cope with the weather and trade changes each year. J. P. Cole and W. V. Tidswell (Hereford College of Education) have both developed games which illustrate this geographical theme. It may also be possible to use the alternatives posed by short-term or long-term conservation of water in dry climates in another simulation of agricultural activity.

The successful operation and management of industrial concerns has already been the subject of several business games, and some of these can be related and adapted for school purposes. (See, for example, the Esso Students' Business Game, developed by Stephen Hargreaves.) These games may also involve useful work in the sphere of economics and/or history.

Locational problems of industry are thoroughly covered in the American High School Geography Project 'Metfab' simulation, previously mentioned, though this is non-competitive.

There are also a number of American games, developed mainly in Social Studies curriculum development projects, which may suggest analogous models useful to British geographers. These include:

(a) EMPIRE—designed by Abt Associates, for the Educational Development Centre, Cambridge, Massachusetts, to demonstrate the place of the American colonists in the British Empire. Students form teams of London merchants, colonial farmers, W. Indies planters, etc., and bargain over goods, move ships across the Atlantic, smuggle(!), etc.

(b) MANCHESTER—also designed by Abt for E.D.C., to show the migration of workers from the country to the city. Students are given roles such as mill-owner, labourer, farmer, etc., and sit round a board which depicts a village with plots of land, a city with factories and a poorhouse.

(c) POLLUTION—designed for the Wellesley Public School System, to demonstrate the problems that face a community as air and water are polluted through the desire to develop industry.

(d) SECTION—developed for the High School Geography Project to demonstrate the conflict of interests between citizens of different sections of a political territory. Students play the roles of citizens of a hypothetical state, comprised of 4 sections —agricultural, prosperous industrial, declining industrial, under-developed rural.

Another useful non-competitive simulation with affinities to game procedures is the model of town-growth developed by M. A. Morgan, and later adapted for classroom use by B. P. FitzGerald and J. A. Everson.

Bibliography

There is little yet written about the use of games in schools, although the subject has produced a quite extensive literature in other professions.

The original stimulus of interest in game *theory* stems largely from J. VON NEUMAN and O. MORGANSTERN's *Theory of games* (Princeton Univ. Press, 1964), but it should be stressed that understanding of game theory is not vital to participation (or even design) in operational gaming.

The conscious use of games in education has, in recent years, begun at university level; one pioneer was H. GUETZKOW of Northwestern University, Evanston, U.S.A., who used several simulation games in teaching international relations; see his chapter in *Game theory and related approaches to social behaviour*, edited by M. SHUBIK (Wiley, 1964). In Britain, comparably, see J. L. TAYLOR and R. N. MADDISON, 'A land-use gaming simulation', *Urban Affairs Quarterly*, June, 1968.

At the school level, there was little until DR J. P. COLE's pamphlet *Geographical games* (University of Nottingham, Dept. of Geography, 1966), in which he illustrated games for small groups of players which he had devised. Some of these were later published in *New ways in geography* by J. P. COLE and N.J. BEYNON, Introductory Book, Books I, II, III and Teachers Book (Blackwell 1969-72).

The following are also of interest:

BERNE, E. *Games people play*, Penguin Books 1968.
CALLOIS, R. *Man, play and games*, Thames & Hudson 1963.
Elementary ideas of game theory, C.A.S. Occasional Paper, H.M.S.O.
FEATHERSTONE, D. F. *War games*, Stanley Paul, 1962.
KLEIN, JOSEPHINE. *Working with groups*, Hutchinson, 1964.
MINSHULL, R. et al *Simulation games in geography*, Macmillan 1972.

VAJDA, s. *Introduction to linear programming and the theory of games*, Methuen 1960.

YOUNG, J. P. *A brief history of war gaming*, Johns Hopkins Univ. Press 1956.

Various publications of 'Project Simile', Western Behavioural Science Institute, 1150 Silverado, La Jolla, California, U.S.A. (Newsletters and simulation exercises of different kinds). See especially 'Plans', 'Napoli' and 'Crisis'.

Eight units of the S.R.A. 'Educational Games Extension Service', written by A. K. GORDON, and published by Science Research Associates, Inc., 259 East Erie Street, Chicago, Illinois.

P. J. TANSEY and D. UNWIN, c/o Bulmershe College of Education, Reading, Berks., have produced valuable book lists and reference material concerned with the general use of simulation techniques in general; they have used them especially in teacher-training. See also their book *Simulation and Gaming in Education*, Methuen, 1969 and *Educational aspects of simulation*, edited by Tansey (McGraw-Hill 1971).

Some commercially marketed games are, of course, of use in the classroom—notably 'Risk' and 'Diplomacy' (Waddington) and many of the war games developed by Avalon-Hill. Their prime aim is entertainment, however, and so sometimes the realities of a situation are considerably changed to suit game excitement.